Rural Tourism and Sustainable Rural Development

Proceedings of the Second International School on Rural Development, 28 June–9 July 1993, University College Galway, Ireland

Edited by

Bill Bramwell and Bernard Lane

CHANNEL VIEW PUBLICATIONS
Clevedon • Philadelphia • Adelaide

Library of Congress Cataloging in Publication Data

International School on Rural Development. Proceedings (2nd; 1993; University College Galway, Ireland)
Rural Tourism and Sustainable Rural Development: Proceedings from the Second International School on Rural Development, 28 June–9 July 1993, University College Galway, Ireland/Edited by Bill Bramwell and Bernard Lane.
Includes bibliographical references.
1. Tourist trade–Congresses. 2. Farms–Recreational use–Congresses.
3. Rural development–Congresses.
I. Bramwell, Bill, 1955- . II Lane, Bernard, 1944- . III. Title.
G155.A1I51.8 1993
338.4'791–dc20 94-21087

British Library Cataloguing in Publication Data

A CIP catalogue record for this book is available from the British Library.

ISBN 1-873150-02-4 (hbk)

Channel View Publications
An imprint of Multilingual Matters Ltd

UK: Frankfurt Lodge, Clevedon Hall, Victoria Road, Clevedon, Avon BS21 7SJ.
USA: 1900 Frost Road, Suite 101, Bristol, PA 19007, USA.
Australia: P.O. Box 6025, 83 Gilles Street, Adelaide, SA 5000, Australia.

 This book is also available as a special issue of the *Journal of Sustainable Tourism*, Vol. 2, Nos 1&2, 1994.

The cover photo shows a traditional sailing vessel, a Galway Hooker, which a community group from Irish-speaking islands plan to use as appropriate visitor transport (© Bernard Lane).

Printed and bound in Great Britain by Short Run Press, Exeter.

CONTENTS

Rural Tourism and Sustainable Rural Tourism

Bill Bramwell
Centre for Tourism, Sheffield Hallam University, UK

Co-Editor: Journal of Sustainable Tourism

This theme issue of the _Journal of Sustainable Tourism_ is devoted to the issues surrounding the development of rural tourism and of more sustainable forms of rural tourism. They are examined in the context of rural areas in the developed nations of Western Europe. Earlier versions of these papers were presented at the Second International School on Rural Development, held in Galway, Ireland in July 1993, which took rural and sustainable rural tourism as its themes. Consideration is given to developing rural tourism in ways where the supply of tourist facilities and experiences is appropriate to the needs of the host community, the environment and the local suppliers, and where it also matches the requirements of tourists on the demand side. Concern is expressed that rural tourism should not develop as the hapless outcome of inexorable, external forces, and hence prominence is given to the role of local communities and local businesses in shaping rural tourism.

Interest in Rural Tourism

It was the pressure of visitors on rural areas in developed nations that led to the development of many of the early ideas about sustainable tourism. Most notably, a number of influential concepts of alternative tourism were developed in the Alpine mountain regions of Europe in the 1970s as an answer to the severe development pressures being put upon rural communities and environments by the rise of intensive winter sports and summer vacationing. It is also the case that many of the new projects and initiatives in sustainable tourism are also to be found in the countryside.

Why have rural areas featured so prominently in the development of sustainable tourism, with its concern for the maintenance of environmental and cultural quality and diversity? A crucial point in many developed countries is the specially high regard in which the countryside is held: it is seen as special and therefore particularly worthy of protection. For example, many observers have noted how in Britain the countryside is highly valued for its inherent variety of landscape, habitat and human heritage (Glyptis, 1991; Middleton, 1982; Prentice, 1992). The cultural commentator Raymond Williams has also shown how in the British literary tradition the countryside has frequently symbolised a lost 'golden age' — an age which supported an idyllic and readily identifiable 'good life'. He illustrates how in this tradition the countryside is defined and experienced as the opposite to life in the town. The countryside is a powerful evocation of a way of life which contrasts favourably with the more utilitarian and materialistic values in urban areas. Rurality embodies all those qualities that are missing from the urban area and modern urban society (Williams, 1975). In many other countries too, the rural areas have become deeply embedded into people's ideas of themselves as a nation (Bowden, 1992). Studies of public attitudes to the countryside also suggest that many people value the countryside because of its symbolic role — the countryside is perceived as a better physical and social world. Whatever the reality of these often idealised values and images, people's views about the qualities of the countryside are often deeply felt (Harrison, 1991). There is nothing new in the high esteem in which the countryside is held, but what is relatively new in many places is a fear for the countryside's future survival in the face of the threat of tourism, with previous threats more likely to be thought to come from industry, mining, depopulation and decline.

The high value which many people (and visitors) place on the countryside provides one justification for this collection of articles. However, the selection of rural tourism as the theme must not be taken to imply that the development of more sustainable tourism is only appropriate for rural areas. Indeed when towns develop tourism then they too must strive to ensure that this development is as sustainable as possible. Urban areas might be more resilient to certain of the adverse pressures of tourism, but their cultures, environments and economies still need careful management and stewardship.

Economic change means many rural areas in Western Europe are facing dilemmas about their future. Changing agricultural practices have led to rural unemployment, and the growth of food surpluses has forced European governments to introduce measures to curtail agricultural production. Many rural areas have suffered from emigration, often of the youngest and most able, and this depopulation has eroded the vitality of rural services and of rural communities. As a result, increased attention has been focused on rural development issues. One particular focus has been to identify ways of encouraging the diversification of rural economic activities, and tourism is one activity receiving heightened attention. Similar trends are evident in developed countries elsewhere. In the United States there has been growing concern about rural economic problems, and a survey in 1991 of state tourism agencies there found that 60% had made increased efforts in rural tourism and 30 of 50 states had tourism programmes specifically targeted for rural areas (Luloff *et al.*, 1994).

Such economic imperatives and the increasing demand for countryside visitation means that many rural areas in Western Europe are under considerable pressure from tourism (FNNPE, 1992). Recognition is growing that tourism can bring many adverse consequences for the countryside, such as the erosion of its distinctiveness. This includes a concern that rural areas should not fall into decline in the tourist-area life cycle (Butler, 1991; Johnson & Snepenger, 1993; Getz, 1992). Rural areas can be under especial threat because of their relative physical fragility, especially in high mountain and other less than robust habitats, and their cultural fragility due to their weakened cultural and political condition after years of economic change and decline, with power flowing to the urban centres. However, the development of sustainable rural tourism has the potential to help secure the economic viability of rural communities and activities, which consequently can help maintain the special qualities of the countryside. The retention of these special qualities is an objective of sustainable tourism more generally, in order to secure long-term conservation and stewardship and also to benefit the host communities and the tourists. Of course rural tourism providers may also see this objective as in their best interests because many tourists who seek out the special characteristics of the countryside might not return if these were lost.

Is Rural Tourism Distinctive?

This collection of papers could be justified by the assertion that rural tourism is relatively distinct from tourism in towns, cities and coastal resorts. The basis of such a claim may appear obvious, but it needs to be demonstrated to be the case. Does the physical existence of tourism in rural areas create a rural tourism that has a significance beyond the self-evident combination of particular activities in a specific place? In other words, do the special characteristics of rural areas help shape the pattern of tourism so that there is a particular rural tourism? This issue is rarely considered in the tourism literature. Consequently, it remains unclear whether rural tourism is distinct from tourism in other geographical settings and, if it is, then in what ways. Here it should be noted that there is a wider debate within the academic discipline of rural studies that questions whether a countryside setting has any significant causal affect on the character of the society and economy of rural areas (Halfacree, 1993). This debate is very pertinent to the question of whether or not there is a distinctive rural tourism.

On one side of the debate in rural studies, there are those who are happy to describe rural areas as possessing to a greater or lesser degree characteristics that distinguish them from urban areas. It is recognised that rural areas are not identical and some are quite similar to urban areas, but it is suggested that to varying degrees rural areas tend to have four main qualities: relatively low physical densities of people, buildings and activities; less social and cultural heterogeneity; less economic diversity; and a comparative physical isolation from general economic, social and political networks. Some researchers have extended these ideas on rural characteristics to suggest that they can have further effects on rural society. Rural society has been depicted as possessing, for example, greater social integration within relatively stable communities (Harper, 1989).

However, such descriptions of rural distinctiveness have been criticised. It is suggested that they are simply based on what we intuitively consider to be rural and that they have not been tested independently from our immediate commonsense thoughts (Newby, 1986). Others argue it is unhelpful to suggest that rural areas tend to have distinctive patterns of social life as there are no necessary connections between a place, its environment and the society found there. Such ideas are reinforced by claims that much of the rural scene in developed countries is becoming progressively less self-contained, due to such influences as television and counter-urbanisation, and has broad access to external values and aspirations (Hoggart, 1990).

Despite these criticisms, some still contend that it may be a mistake to deny our commonsense thoughts that rural areas can have distinctive characteristics or that these can have consequences for social and economic interactions in the countryside. Certainly, people do tend to hold images and views of the countryside that differ from those of the town; in this sense the countryside undoubtedly is distinctive, and this distinctiveness has at least two inextricably related dimensions. There is both our individual responses to the countryside and also the popular social representations of the countryside, such as those in the media, with the latter reflecting and affecting our individual perspectives. These social representations and personal responses to the countryside are important, not least because they affect how people interact with rural areas. In terms of rural tourism, they affect whether people want to visit the countryside and what they see and do when they are there (Squire, 1994).

Whatever the extent to which rural tourism is a distinct activity, it is hoped that this collection of papers presents some useful insights into issues which concern those interested in tourism in rural areas. Some of these issues are presented elsewhere (in Bouquet & Winter, 1987; Gilbert, 1989; and Pigram, 1992) and these are related to case studies in Bramwell (1990) and in Hawkes & Williams (1993). This collection might also be of value to those concerned more generally with the theory and practice of sustainable tourism.

Sustainable Rural Tourism and Tourism Demand and Supply

Several of the articles here examine tourism in rural areas in relation to the matching of the supply of tourist facilities and experiences to the character of the demand for tourism. They assess how tourism in the countryside could be more sustainable when it meets the requirements of the local community and environment and of local suppliers on the supply side of the tourism system as well as of the tourists on the demand side.

Richard Hummelbrunner's article suggests that a combination of changes on the demand and supply sides of tourism has led to the emergence of a 'new rural tourism'. The supply side changes include an increasing interest in rural tourism among a better-off clientele, and also among some holidaymakers a growing environmental awareness and a desire to be integrated with the residents in the areas they visit. Hummelbrunner considers these trends are likely to promote forms of rural tourism supply which build on the inherent character of rural areas, notably their attractive natural environments, original local culture and traditional systems of landuse and farming. In his view, rural areas should look inward to their own resources when developing rural tourism, although the common rural weaknesses of a lack of local entrepreneurs and of local capital and expertise may mean that at times it is sensible to seek external or exogenous capital. It is important, however, that this external involvement is managed with caution. Such deficiencies as a lack of local entrepreneurs can be tackled by establishing a network of advisors to build on endogenous entrepreneurial resources. He illustrates this by examining the local advisory and consultancy support provided by the

Austrian Association for Regional Self-Reliant Development.

Agnes Gannon's article contends that more attention often needs to be paid in rural tourism to the needs of the consumer, that is, to the requirements of tourism demand. This can require a shift from 'selling what we can produce' to 'producing what we can sell'. It must be balanced, however, by attention to the needs of the rural locality, including striving for authenticity, involving the community, and promoting human values and ecological stability.

Changes in rural tourism demand are also seen as important by Xavier Greffe. Greffe considers that the shift in the socio-economic profile of those interested in rural tourism towards the better-off socio-economic groups suits a move from the common, but often self-defeating, search for economies of scale in tourism to the search for economies of scope, that is, from maximising the number of tourists towards an emphasis on increasing the opportunities for them to spend. This emphasis on economies of scope may be very relevant to the development of sustainable tourism in rural areas. He also considers some of the requirements for tourism supply in rural areas given this preference for economies of scope. Prominent among these are a need for government to assist in the renovation and extension of the supply conditions in order to attract and hold on to the strong potential demand from better-off visitors.

Another requirement identified by Greffe is for rural tourism supply to be organised in networks which link together individual suppliers. This is illustrated in the article on agri-tourist businesses in Austria by Hans Embacher. He argues that, while the small scale of farm tourism businesses is a strength in relation to sustainable tourism, this is also a weakness in the market. For these small operators to reach the market effectively there needs to be co-operative marketing. Embacher examines the demand and supply characteristics of agri-tourism in Austria as well as the assistance which farm tourism can provide to sustaining rural life in general and the farming sector in particular. Detailed consideration is also given to the Austrian Farm Holidays Association as a co-operative marketing organisation which links individual farm businesses at regional, provincial and federal scales and which provides a range of marketing, training, advice and educational services to its members.

Concepts and Methods of Sustainable Rural Tourism

Lars Aronsson's paper on sustainable rural tourism in Sweden directs its attention to conceptual frameworks which link tourism demand and supply to sustainable tourism. He advocates the use of the concept of the 'meeting place' as a way of examining the encounters between tourist demand (in the sense of people's values and consumer patterns) and the tourist product, including the environment and society of tourist destinations. Proposals are put forward for what may be required of a tourism system — with its two main aspects, demand and supply, and the 'meeting places' between them — to be sustainable. The other article which highlights conceptual frameworks is by David Crouch. He argues that visitors' perceptions of rural places and cultures and their 'demand' for tourism is shaped in part by the tourist promotion. Taking rural Ireland as a case study, he concludes that the tourist promotion of Ireland caricatures the rural cultures found there. For Crouch, a central task in developing sustainable tourism is the construction of a different set of cultural 'identities', and these should present rural cultures very differently from how they are presented when simply directed at the tourist consumer. This concept of 'identities' is developed from recent ideas about culture in post-modern society. A central concern in the shaping of these 'identities' is the need to give people in rural communities the confidence to draw upon their own experiences and expertise; in other words, the need for community tourism and empowerment.

The last two articles examine the practical application of methods which are potentially of use in the development of sustainable rural tourism. Bernard Lane assesses sustainable tourism strategies as a method to achieve both development and conservation in rural areas. The paper by Adrian Phillips and Mary Tubridy explains how they have applied a range of methods to assist in the development of rural tourism in the Irish counties of Wicklow and Wexford, with the intention that this tourism is stably rooted in the rural economy, enhances the well being of the community, and draws on the unique local natural and human heritage. The methods which they have used and are examined here include the production of resource databases using Geographical Information Systems, the development of interpretive plans, as well as methods of involving rural communities in interpreting their heritage.

Conclusion

Among the differing contributions on rural tourism in this collection, there is a frequent concern that rural tourism should not be allowed to develop as the inevitable outcome of powerful and inexorable outside forces. These forces include the shifting and increasingly global pressures for economic restructuring and the commoditisation of the countryside by external tourism interests as a retailed experience for tourist consumers. A new concern is emerging in the theory and practice of rural tourism which invests considerable importance in the economic well-being and cultural integrity of rural communities and in rural environmental stewardship. Increasing prominence is now given to rural communities and individual local residents and businesses and to their role and degree of control in shaping rural tourism in the context of external forces (Britton, 1991; Drake, 1991; Evans & Ibery, 1989; Getz, 1994; Long *et al.*, 1990; Marsden, 1992; Prentice, 1993). Inevitably there will be contested opinions on what is desirable in the development of rural tourism in specific places. It is hoped that these articles will help clarify what may be desirable, indicate possible ways forward, and also stimulate further analysis and debate.

References

Bowden, M. (1992) The invention of American tradition. *Journal of Historical Geography* 18, 3–26.

Bouquet, M. and Winter, M. (eds) (1987) *Who From Their Labours Rest? Conflict and Practice in Rural Tourism.* Aldershot: Gower.

Bramwell, B. (ed) (1990) *Shades of Green: Working Towards Green Tourism in the Countryside.* London: English Tourist Board, Countryside Commission.

Britton, S. (1991) Tourism, capital and place: Towards a critical geography of tourism. *Environment and Planning D* 9, 451–78.

Butler, R. (1991) Tourism, environment and development. *Environmental Conservation 18 (3), 201–9.*

Drake, S.P. (1991) Local participation in ecotourism projects. In T. Whelan (ed.) *Nature Tourism: Managing for the Environment* (pp.132–63). Washington: Island Press.

Evans, N.J. and Ibery, B.W. (1989) A conceptual framework for investigating farm-based accommodation and tourism in Britain. *Journal of Rural Studies* 5 (3), 257–66.

FNNPE (The European Federation of Nature and National Parks) (1992) *Loving Them to Death? The Need for Sustainable Tourism in Europe's Nature and National Parks* (Final Report to the European Commission). Brussels: European Commission.

Getz, D. (1992) Tourism planning and destination life cycle. *Annals of Tourism Research* 19, 752–70.

— (1994) The environment–community symbiosis: A case for collaborative tourism planning. *Journal of Sustainable Tourism* 2 (3), forthcoming.

Gilbert, D. (1989) Rural tourism and marketing: Synthesis and new ways of working. *Tourism Management*, March, 39–50.

Glyptis, S. (1991) *Countryside Recreation.* London: Longman.

Halfacree, K.H. (1993) Locality and social representation: Space, discourse and alternative definitions of the rural. *Journal of Rural Studies* 9 (1), 23–37.

Harper, S. (1989) The British rural community: An overview of perspectives. *Journal of Rural Studies* 5, 161–84.

Harrison, C. (1991) *Countryside Recreation in a Changing Society.* London: TMS Partnership.

Hawkes, S. and Williams, P. (eds) (1993) *From Principles to Practice. A Casebook of Best Environmental Practice in Tourism.* Burnaby: Centre for Tourism Policy and Research, Simon Fraser University.

Hoggart, K. (1990) Let's do away with rural. *Journal of Rural Studies* 6 (3), 245–57.

Johnson, J.D. and Snepenger, D.J. (1993) Application of the tourism life cycle concept in the Greater Yellowstone Region. *Society and Natural Resources* 6, 127–48.

Long, P.T., Perdue, R.R., Allen, L. (1990) Rural resident tourism perceptions and attitudes by community level of tour-

ism. *Journal of Travel Research*, Winter 1990, 3–9.
Luloff, A., Bridger, J., Graefe, A., Saylov, M., Martin, K. and Gitelson, R. (1994) Assessing rural tourism efforts in the United States. *Annals of Tourism Research* 21, 46–64.
Marsden, T. (1992) Exploring rural sociology for the Fordist transition: Incorporating social relations into economic restructuring. *Sociologia Ruralis* 32, 209–30.
Middleton, V. (1982) Tourism in rural areas. *Tourism Management* 3 (1), 52–8.
Newby, H. (1986) Locality and rurality: The restructuring of rural social relations. *Regional Studies* 20, 209–15.
Pigram, J.J. (1992) Planning for tourism in rural areas: Bridging the policy implementation gap. In D.G. Pearce and R.W. Butler (eds) *Tourism Research: Critiques and Challenges* (pp. 156–74). London: Routledge.
Prentice, R. (1992) The Manx National Glens as treasured landscape. *Scottish Geographical Magazine* 108 (2), 119–27.
— (1993) Community-driven tourism planning and residents' preferences. *Tourism Management*, 14, 218–27.
Squire, S.J. (1994) The cultural values of literary tourism. *Annals of Tourism Research* 21, 103–20.
Williams, R. (1975) *The Country and the City*. London: Paladin.

What is Rural Tourism?

Bernard Lane
Rural Tourism Unit, Department for Continuing Education, University of Bristol, Bristol BS8 1HH, UK

This paper reviews the development of tourism in rural areas. It defines rural tourism as a discrete activity with distinct characteristics which may vary in intensity, and by area. It discusses the differences between agri-tourism and rural tourism, and examines why there should be a special relationship between tourism in the countryside and the concept of sustainable tourism.

Across the rural regions of the developed world the issues of population decline, economic change and community regeneration are almost universal. For over a century, the powerful trends of industrialisation and urbanisation have steadily altered the economic and political position of rural society. In the last forty years those trends have intensified. Farm incomes have fallen in real terms and technological changes have joined with falling incomes to reduce agricultural employment. In response, rural service provision has diminished: shops, schools, churches, professional services and transport facilities have all declined in numbers and in underlying vitality. Typically, rural populations have aged and become fewer in total. Many small towns and villages now struggle to retain their viability. Everywhere, local, regional and central government agencies have intervened to address these issues, with various degrees of success.

In recent years, the rural world has seen new challenges. Nature and landscape conservation is increasingly regarded as important. Historic buildings and 'traditional' rural societies are receiving more attention. In some more accessible rural regions, there has been an influx of population, often of people unhappy about big city living conditions — a trend known as counter-ubanisation. But for many parts of the countryside, rural decline and development issues remain important (Pacione, 1984; Robinson, 1990; OECD, 1993).

In contrast to the picture of decline and change in agriculture and in the countryside, the world of tourism has experienced rapid expansion since the early 1950s. International tourist arrivals have increased from 25 millions in 1950 to an estimated 476 millions in 1992. The World Tourism Organization estimates domestic travel numbers to be ten times the international total: it is also growing rapidly. The WTO estimates that, by 2000, tourism could be the world's largest single industry (WTO; Edgell, 1993).

There are many factors behind the growth statistics. These include rising disposable incomes, longer paid holidays, shorter working hours, better education and rapidly improving transport and communications networks Rarely, however, have government agencies directly influenced the pace of tourism development, which has largely been driven by market forces and private enterprise. In this way, tourism differs markedly from agriculture and other rural activities.

Tourism was, and still largely is, concentrated into specialist beach, lake and mountain resort areas, and into major cultural centres. It has proved to be a powerful engine for economic growth — transferring capital, income and employment from industrial, urban and developed areas to non-industrial regions. A study of the growth of tourism in, for example, post-war Spain, Greece, Austria or even Ireland is a revealing commentary on the economic power of the leisure industries (Williams & Shaw, 1988). But, in recent years, the rise of the independent traveller, seeking a non-resort-based or rural holiday, has now added a new dimension to the travel industry (see Lane, 1990; Plog, 1991; Weiler & Hall, 1992). This trend towards holidays in the countryside has considerable potential as a tool for economic regeneration. That is one of the major reasons why administrations which have rural remits have begun to develop an interest in tourism development.

This paper examines a number of basic issues which are important for administrators, politicians and entrepreneurs who wish to promote tourism in the countryside. It seeks to define rural tourism in the developed countries of the world, both in terms of its characteristics and in terms of the types of holidays and holiday experiences offered. It goes on to discuss the relationship between agri-tourism and rural tourism, and concludes by examining why there is a special need for rural tourism to embrace sustainable tourism.

The Development of Rural Tourism

Rural tourism is not totally new. Interest in countryside recreation grew in the nineteenth century as a reaction to the stress and squalor of the expanding industrial cities. Writers such as Wordsworth and Schiller captured the romanticism of the rural scene. The new railway companies capitalised on this emergent interest by transporting tourists to the countryside. The Alps and the American and Canadian Rockies were early rural tourism venues assisted by rail-led marketing and capital investment (Feifer, 1985; Runte, 1990).

The new rural tourism of the 1970s, 80s and 90s is, however, different in several ways. Far larger numbers of visitors are involved. Over 70% of all Americans now participate in rural recreation: figures for many other developed countries reveal similar, if slightly lower, levels of participation. The penetration of those visitors is far greater. The spread of car ownership and internationally available car hire allows visitation far beyond rail-nets and rail-heads. Most important of all, tourism has developed away from spectacularly scenic areas into countryside of all types. It has also broken free of large and specialised resorts into small towns and villages to become truly rural (Knudson, 1984; Bramwell, 1990).

Growth in rural tourism is difficult to quantify, because few countries collect statistics in a way which separates purely rural from other forms of tourism. Most national tourism administrations agree, however, that it is a growth sector. Individual rural regions provide further testimony. In the American states of Wyoming, Montana and Idaho, ranch and farm hospitality enterprises have increased from five in 1985 to ninety in 1992 (Bryan, 1991). Australia's 'Outback Queensland' area reports steady annual increases in visitor flows of up to 20% pa (Hall, 1991). Many new tourism projects in rural Europe have achieved a steady development in patronage throughout the 1980s (Keane & Quinn, 1990; Greffe, 1992).

But What is Rural Tourism?

At first glance, this is a simple question. Rural tourism is tourism which takes place in the countryside. But, as many early commentators have pointed out (e.g. Keller, 1990; Greffe, 1992), on deeper consideration, a simple definition of rural tourism is inadequate for many purposes. Equally, it is difficult to produce a more complex definition which applies to all rural areas in all countries. Problems include:

(1) Urban or resort-based tourism is not confined to urban areas, but spills out into rural areas.
(2) Rural areas themselves are difficult to define, and the criteria used by different nations vary considerably.
(3) Not all tourism which takes place in rural areas is strictly 'rural' — it can be 'urban' in form, and merely be located in a rural area. Many so-called holiday villages are of this type; in recent years, numerous large holiday complexes have been completed in the countryside. They may be 'theme parks', time shares, or leisure hotel developments. Their degree of rurality can be both an emotive and a technical question (see Murdoch, 1993; CPRE, 1994).
(4) Historically, tourism has been an urban concept; the great majority of tourists live in urban areas. Tourism can be an urbanising influence on rural areas, encouraging cultural and economic change, and new construction.
(5) Different forms of rural tourism have developed in different regions. Farm-based holidays are important in many parts of rural Germany and Austria. Farm-based holidays are much rarer in rural USA and Canada. In France, the self-catering cottage, or gîte, is an important component of the rural tourism product.
(6) Rural areas themselves are in a complex process of change. The impact of global markets, communications and telecommunication have changed market conditions and orientations for traditional products. The rise of environmentalism has led to increasing control by 'outsiders' over land use and resource development. Although some rural areas still experience depopulation, others are experiencing an inflow of people to retire or to develop new 'non-traditional' businesses. The once clear distinction between urban and rural is now blurred by suburbanisation, long distance commuting and second-home development (see Robinson, 1990).
(7) Rural tourism is a complex multi-faceted activity: it is not just farm-based tourism. It includes farm-based holidays but also comprises special-interest nature holidays and ecotourism, walking, climbing and riding holidays, adventure, sport and health tourism, hunting and angling, educational travel, arts and heritage tourism, and, in some areas, ethnic tourism. There is also a large general-interest market for less specialised forms of rural tourism. This area is highlighted by studies of the important German tourism market, where a major requirement of the main holiday is the ability to provide peace, quiet and relaxation in rural surroundings (Studienkreis fuer Tourismus, 1987).

Because rural tourism is multi-faceted, because rural areas themselves are multi-faceted and rarely either static entities or self-contained, or free from urban influence — because of all these points, a working and reasonably universal definition of the subject is difficult to find. However, in almost every case rurality is the central and unique selling point in the rural tourism package. The search for a definition must, therefore, begin with an understanding of the concept of rurality itself.

What is Rurality?

The need for a definition of rural tourism is relatively new. The need for a definition for rurality in general is an old issue faced by geographers, sociologists, economists and planners for many years. Much of their experience can be used to help produce a better working concept for tourism in the countryside (see OECD, 1992; Robinson, 1990; Carter, 1981). In this wider debate on rurality three major discussion points dominate: (1) population density and size of settlements; (2) land use and economy, and the dominance of agriculture and forestry; (3) 'traditional' social structures and issues of community identity and heritage. It is worth briefly discussing each of these in turn.

Rural population densities and settlement size

Typically rural areas have low population densities: this is a result of small settlements, widely spaced apart. The natural and/or the farmed/forested environment dominates the built environment. Average rural population densities vary enormously between and within developed countries: an exact analysis would be valueless because of the varying sizes of the administrative units used for statistical purposes. This point is also illustrated when examining the size of settlements classified as rural by a selection of countries, given below.

Selected national criteria for 'rural' settlements

Australia	Population clusters of less than 1,000 people, excluding certain areas, e.g. holiday resorts.
Austria	Parishes of less than 5,000 people.
Canada	Places of less than 1,000 people, with a population density of less than 400 per square km.
Denmark (and Norway)	Agglomerations of less than 200 inhabitants.
England and Wales	No definition — but the Rural Development Commission excludes towns with more than 10,000 inhabitants.
France	Parishes containing an agglomeration of less than 2,000 people living in contiguous houses, or with not more than 200 metres between the houses.
Portugal (and Switzerland)	Parishes of less than 10,000 people.

Source: *UN Demographic Year-books* and Robinson (1991)

The OECD Rural Development Programme uses a pragmatically based series of indicators: while at local level a population density of 150 persons per square kilometre is the preferred criteria, 'at the regional level geographic units are grouped by the share of their population which is rural, into the following three types: predominantly rural (>50%), significantly rural (15–50%) and predominantly urbanised regions (<15%)' (OECD, 1993).

From this array of varying definitions, two clear points stand out. Rural settlements may vary in size, but they are small, and always with a population of less than 10,000 inhabitants. They are almost always in areas of relatively low population density. Both of these points are important in tourism terms: the majority of tourists come from densely populated and large settlements — they seek a change of scene while on holiday. Tourism is in many respects a form of escapism (Plog,1991).

Land use and economy

'Rural land encompasses areas which are under agriculture, forest and woodland, as well as wild uncultivated tracts in a natural or semi-natural state' (Best & Rogers, 1973). More precisely, many commentators define rural areas as those with less than 10–20% of their land areas covered by the built environment. There are three important implications here. These areas will be visually dominated by agrarian and forest-based economic activities. They will be, to a large extent, repositories of the natural world and wildlife. For the visitor, they will give an impression of space, and a traditional non-urban, non-industrial economy. Their economies will be strongly influenced by the market for farm and forest products. Although the labour force required for farming and forestry has declined rapidly in recent years, rural areas still show a strong bias towards jobs in the farm/forest sector. Additionally, they usually exhibit low female activity rates, because of the shortage of job opportunities for women in many rural areas. The tourism industry has traditionally been an important employer of female labour. Rural tourism has, therefore, a special importance for rural women seeking to enter employment or self-employment, as well as other impacts on the sociology of the rural family (Bouquet & Winter, 1987)

'Traditional' social structures

The rapid urbanisation of the nineteenth and twentieth century produced new social structures different from the 'traditional' societies of the countryside. The retention of older ways of life and thinking is important in retaining rural 'character'. It is this residual character which, combined with the scenic values and recreational opportunities of the countryside, attracts tourists from urban areas. The rural heritage 'product' includes non-physical heritage as well as tangible features from the man-made and natural world.

It is difficult to define the exact characteristics of rural society. There are great variations between countries and continents, and even within countries. Flinn (1982) noted three very different types of traditional life styles within rural USA:

(1) Small-town society, closely knit, strongly believing in democracy, but often not in close contact with nature.
(2) Agrarian society, based on family farming, farm life and the calendar of the seasons.

(3) Ruralists, living outside towns, but not farming: independents who value open space, nature, and 'a natural order'.

The rural sociologists have struggled hard to identify the varying characteristics of rural societies. Frankenberg's urban/rural contrasts, dating from 1966, remains a valuable check list (see Table 1).

Table 1 The characteristics of 'rural' and 'urban' societies

Rural	*Urban*
1. Community	1. Association
2. Social fields involving few but multiple role relationships	2. Social fields involving many overlapping role relationships
3. Different social roles played by same person	3. Different social roles played by different people
4. Simple economies	4. Diverse economies
5. Little division of labour	5. Great specialization in labour force
6. Ascribed status	6. Achieved status
7. Education according to status	7. Status derived from education
8. Role embracement	8. Role commitment
9. Close-knit networks	9. Loose-knit networks
10. Locals	10. Cosmopolitans
11. Economic class is one of several divisions	11. Economic class is the major division
12. Conjunction	12. Segregation
13. Integration with work environment	13. Separation of work environment

Source: Frankenburg, 1966

The Continuum Concept

Faced with the complexities of the rural world discussed above, and with an array of other indices deliberately not discussed here, commentators on the rural scene have evolved the concept of the rural–urban continuum as a way of coping with the complexity of the situation, and the problem of comparing areas which are perceived to be rural, but possess many different characteristics. Rural communities can be assessed on a sliding scale with sparsely populated remote wilderness as one end of a polar typology. The other end of the scale can be represented by the so-called 'world city', the ultimate expression of urbanisation. Between these extremes lie a variety of situations, largely rural or largely urban, with a mid-point represented by the outermost edge of suburbia, a cross-over point between poles, exhibiting characteristics of both rural and urban typology (see Robinson, 1990).

An additional part of the rurality equation can be introduced here. The OECD's Rural Development Programme has developed a useful typology for assessing the economic geography of rural areas (OECD, 1992). This divides the rural world into peripheral or remote regions, 'intermediate' regions, which make up the majority of the rural land mass, and economically integrated rural regions, often close to large urban complexes. The peripheral regions are characterised by sparse populations, small-scale often traditional enterprises, high servicing costs and economic poverty. The economically integrated regions tend to have large farm units, a diversified economy, good services and relative affluence. The intermediate areas lie midway between these extremes. This typology can also fit the continuum concept: the three types of region are not sharply defined but blend into each other.

The OECD's typology has important implications for both rural development and rural tourism. In tourism terms, one critical point is that economically integrated rural regions tend to develop a strong day-visitor trade, and may experience high levels of visitation. Both present specialised commercial opportunities, but also present visitor management problems. In contrast, day visitors and high levels of visitation are rare in peripheral regions, although visitor management is still important, in order to obtain a flow of repeat business, and to safeguard the special cultural qualities which may be found in peripheral regions.

The continuum concept not only copes with a variety of landscapes, life styles and demographic inheritance — it also copes with change. Settlements can move along the continuum, exhibiting change through time. Typically — but not always — the change tends to move districts and settlements towards the urban pole. On the OECD's economic typology, change is also common — and can also occur in either direction.

A further valuable point to note is that the continuum concept allows observers to realise that indices of rurality may change at different rates relative to each other. Thus, rural settlements may remain locationally rural while becoming functionally urban. Settlement size and population density may remain 'rural', but economies may become non-agricultural, and society may display increasing numbers of urban characteristics. Different parts of rural society may display more or less urban characteristics. The American geographer Brian Berry notes the increasing development in the USA of 'urban civilisation without cities' as educated, independently employed or retired urbanites seek 'rustic backgrounds for sophisticated lives lived in a country setting' (Berry, 1980). Commentators have, however, also noted that these newcomers to the countryside are amongst the most vocal in seeking to preserve certain rural characteristics, notably, landscapes dominated by farming, forestry or parkland, and small-scale settlements.

A final comment in this discussion is necessary. Throughout the debate on definitions of rurality amongst academics, planners and politicians, rurality is almost always seen as an important condition, possessing very valuable characteristics worthy of preservation. Thus, while the global condition has become steadily more urban, most commentators have stressed the importance of retaining key differences between urbanisation and the rural realm. Forty years ago, in 1951, the British government set up a special commission to review the future of that most remote and most rural part of Britain, the crofting communities of the Highlands and Islands of Scotland. (Crofting is a subsistence farming/fishing economy carried out on smallholdings, usually of less than five hectares, far away from urban settlements.) After noting that crofting flew in the face of all modern agrarian practices, and required, even in 1951, massive levels of state subsidy, the Commission concluded that, 'in the national interest, the maintenance of these communities is desirable, because they embody a free and independent way of life which, in a civilisation predominantly urban and industrial in character, is worth preserving for its own intrinsic value' (HMSO, 1951). In 1990, Peter Keller, in a report for the Tourism Committee of the OECD, on 'Tourism Policy and Rural Development', wrote 'The countryside as a creative counterbalance to the hypercivilised urban centre is no illusion...hill farmers must be kept on the land...highly developed economies should be able to afford the luxury of safeguarding typically rural areas' (Keller, 1990). It is important to note that Peter Keller is not a rural sociologist or a representative of the farming community: on the contrary, he is an expert working in the tourism industry. Nevertheless, he too recognised the special value of the countryside.

The practical expression of the pro-rural sentiments quoted above has been two-fold. Government policies financially support both primary rural products and rural service provision in every country of the developed world. These financial policies are frequently backed by strong conservation policies, aimed at the retention of existing landscape areas, whole settlements, buildings and the natural world. Urban conservation policies also exist, but usually cover only fragments of cities and isolated special situations.

Rural Tourism: A Special Concept?

It has been argued above that rurality as a concept is connected with low population densities and open space, and with small-scale settlements, generally of less than 10,000 inhabitants. Land use is dominated by farming, forestry and natural areas. Societies tend towards traditionalism: the influence of the past is often strong. Government policies lean towards conservation rather than radical or rapid change.

It may be argued, therefore, that rural tourism, in its 'purest' form, should be:

(1) Located in rural areas.
(2) Functionally rural — built upon the rural world's special features of small-scale enterprise, open space, contact with nature and the natural world, heritage, 'traditional' societies and 'traditional' practices.
(3) Rural in scale — both in terms of buildings and settlements — and, therefore, usually small-scale.
(4) Traditional in character, growing slowly and organically, and connected with local families. It will often be very largely controlled locally and developed for the long-term good of the area.
(5) Of many different kinds, representing the complex pattern of rural environment, economy, history and location.

A list of contrasting features between urban/resort tourism and rural tourism is given in Table 2.

It is interesting to note that many of the features noted in the left-hand or urban/resort column were once thought to be vitally necessary for the successful

Table 2

Urban/Resort Tourism	Rural Tourism
Little open space	Much open space
Settlements over 10,000	Settlements under 10,000
Densely populated	Sparsely populated
Built environment	Natural environment
Many indoor activities	Many outdoor activities
Infrastructure — intensive	Infrastructure — weak
Strong entertainment/retail base	Strong individual activity base
Large establishments	Small establishments
Nationally/internationally owned firms	Locally owned businesses
Much full time involvement in tourism	Much part-time involvement in tourism
No farm/forestry involvement	Some farm/forestry involvement
Tourism interests self supporting	Tourism supports other interests
Workers may live far from workplace	Workers often live close to workplace
Rarely influenced by seasonal factors	Often influenced by seasonal factors
Many guests	Few guests
Guest relationships anonymous	Guest relationships personal
Professional management	Amateur management
Cosmopolitan in atmosphere	Local in atmosphere
Many modern buildings	Many older buildings
Development/growth ethic	Conservation/limits to growth ethic
General in appeal	Specialist appeal
Broad marketing operation	Niche marketing

development of a tourist industry. They may now be less important. Densely populated resorts, in an international style, under multi-national ownership, and with a relatively anonymous management may actually deter some types of client. In contrast, the countryside's open spaces, lack of indoor facilities, locally owned and sometimes unprofessionally managed businesses, and older buildings may be a main selling point for rural holidays (Plog, 1991). Ironically, however, some rural communities see the development of resort facilities as a key to their future success in the world of tourism.

The Importance of the Continuum Concept for Rural Tourism

Commentators seeking to define rurality have made extensive use of the concept of the rural/urban continuum to deal with many different types of area, exhibiting different characteristics, and areas undergoing active change. A similar continuum concept can be useful for those seeking to define rural tourism. Few areas will display all of the characteristics of rural tourism listed previously. Many will display some 'urban' characteristics. Some will be in the process of change and development towards becoming large, urban resorts. The use of the continuum concept allows planners to recognise this trend, and to take steps either to regulate it, or to make infrastructural provision for it. It can be strongly argued that management strategies in rural tourism should aim to conserve rurality as an important resource. But, in some cases it may be valuable to allow or even encourage some change to take place.

What Types of Holidays are Rural?

This is a frequently asked and extremely difficult question. Rural tourism cannot be defined solely by holiday type: intensity of use, location, style of management, integration with the community and other factors play an important part in the definition. But a broad-brush approach can be useful. Again, the continuum concept is a useful one. Many types of holiday can be developed in both urban and rural locations. Holiday-makers may be involved in both urban and rural activities on the same day. A tentative classification of holiday types and activities is given below: it should be used with care. The listing follows the continuum concept, moving from specifically rural to specifically urban with a broad intermediate category.

Holiday activities which are usually specifically rural

1. Walking
2. Climbing
3. 'Adventure' holidays/wilderness holidays
4. Canoeing
5. Rafting
6. Cross-country skiing
7. Snow-shoe tours
8. Low intensity downhill skiing
9. Nature study in outdoor settings, including birdwatching, photography etc.
10. Hunting
11. Cycling/cycle touring
12. Horse riding
13. Landscape appreciation.
14. Rural heritage studies
15. Small town/village touring
16. Relaxation holidays requiring a rural milieu
17. Small-scale conventions/conferences
18. Rural festivals
19. River and canal angling
20. Sports requiring natural settings, e.g. orienteering

Holiday activities which may be rural or urban/resort based

1. Swimming
2. Low/medium intensity beach holidays
3. Medium intensity downhill skiing
4. Sports requiring man-made infrastructure of a semi-natural type, e.g. golf
5. Cuisine-based holidays

6. General heritage holidays
7. Conservation holidays
8. Educational holidays
9. Cultural festivals
10. Craft holidays
11. Camping
12. Sightseeing/touring
13. Small/medium-sized conferences/conventions
14. Sailing/cruising
15. Sea angling

Holiday activities which are usually specifically urban/resort based

1. City sightseeing
2. Shopping
3. High intensity beach holidays
4. High intensity downhill skiing
5. Urban heritage/culture holidays
6. Zoological gardens
7. Health resorts
8. Industrial tourism
9. Major conferences/conventions
10. Entertainment holidays/gambling
11. Resort holidays
12. Sports requiring man-made infrastructure, e.g. international arena-based events

The Relationship between Tourism and Agriculture

Traditionally, agriculture and forestry were central to rural life. They were the major employers of labour, the main sources of income within the rural economy, and indirectly had a powerful influence on traditions, power structures and life styles. Together, the decisions of farmers and foresters determined rural land use and landscapes.

In the late twentieth century, the central role of farming and forestry has been diminished. Both activities have shed much of their labour force. Only five OECD countries now employ more than 15% of their labour force in farming, forestry and fishing: in eight OECD countries, that figure is less than 5% (OECD, 1993). The economic power of farming and forestry has declined, not least, because those activities are extremely dependent on state subsidies for their profitability. Traditions are waning before the combined attack of television, 'power-farming' techniques and tree processing machinery. Even the role of agri- and arboriculturalists as 'landscape gardeners' has diminished: power has begun to move to planners and conservationists.

In this evolving situation, two myths have grown up about the role of rural tourism. One is that rural tourism is farm-based, or agri-tourism. The second is that diversification into tourism will 'save' the farming community. These statements are not universally true because the relationships between agriculture, forestry and tourism are extremely complex ones.

Farm-based rural tourism has been successful through many (but not all) parts of German-speaking Europe because of a powerful combination of small farm size, interesting scenery, closeness to markets, traditional town/country links caused by late migration from the countryside to city regions, the owner-occupation of farmsteads and the tradition of effective and interventionary local government and co-operative movements. Furthermore, the role of farm-based tourism has been exaggerated because it has received great attention from both agriculture ministries and academics. A bibliography of rural tourism for the OECD (Lane, 1993), revealed that farm tourism is the largest single special category of rural tourism in terms of published works. In areas where some or all of the factors mentioned above are lacking, farm-based tourism has been slow to develop. Reasons include:

(1) Long distances to the urban holiday market.
(2) Medium and large sized farms which did not need to diversify.
(3) Rented farms which either failed to receive the owner's permission to diversify, or were amalgamated to create larger units.
(4) Very poor and very small farms which had no surplus accommodation.
(5) Co-operatives, local councils and tourist boards which did not help with marketing and infrastructure provision.
(6) Scenery/heritage/activity attractions which were poor.
(7) A short, single season.

Thus, for example, over large parts of Eastern England, Sweden, Canada and USA, farm tourism is poorly developed. But that does not mean that rural tourism is poorly developed. Many of the kinds of rural holidays discussed earlier are not dependent on farm situations. Accommodation can be provided by hotels and motels, small town and village bed and breakfasts, purpose built lodges, camping and caravan sites. And in areas where agri-tourism is established, it may form only a part of the rural tourism product, which embraces the totality of the rural world.

Diversification into rural tourism is frequently held up to be a potential panacea for agriculture's ills. There is no doubt that in some areas, and for some businesses, tourism can be valuable. But there are serious problems in its universal application:

(1) Over 75% of the land of the developed world is rural: there are insufficient visitors to maintain all farmers in all areas.
(2) Farm-based tourism may not reduce productivity on many farms. Additional tourism earnings may be invested to increase agricultural productivity. Therefore, farm surpluses could continue to grow, leading to falling prices and quota restrictions, and a further round of farm problems (Denman & Denman, 1990; ADAS, 1992).
(3) Some areas are unsuitable for intensive tourism development because of remoteness, lack of scenic or heritage attractions, and other factors mentioned earlier in this section.
(4) Successful farm tourism development seems to require effective co-operative marketing and development efforts. Many areas have no tradition of co-operation between farmers, or between farmers and governmental agencies.

Rural Tourism and Sustainable Tourism

Tourism was once promoted as a clean and harmless economic activity, free of the environmental impacts attributed to manufacturing, mining, logging and intensive agri-business. Since the 1970s, this benign view of tourism has been increasingly questioned. The physical impact of feet, litter and vehicles, the economic impact of tourism's multi-nationals, the social impact of mass visitation on cultures, heritage areas and objects — all — and more — are now seen as problems. The discussions about the down side of tourism are well documented (see for example, Krippendorf, 1987; English Tourist Board, 1991). From these discussions came the now well known concept of sustainable tourism, an approach to managing tourism which tries to minimise its harmful impacts while maximising its benefits (Bramwell & Lane, 1993).

The countryside is especially at risk from unmanaged — or ill-managed — tourism. Parts of the physical environment are at risk because they are fragile: natural habitats, archaeological features and footpaths are but a few of the features under threat from visitor numbers. The economic stability of the rural world can be at risk from new, perhaps large scale, tourism businesses seeking short-term gains. And the cultural heritage may be at risk from outsiders and from powerful outside cultures. For all these reasons, a sustainable approach to rural tourism would seem to be especially important. These are, however, problems which mass tourism may bring to many places, urban or rural, resort area or a conventional settlement.

There are four additional reasons why rural tourism must consider a sustainable form of tourism management if it is to be successful in the long term. The first of these reasons is that visitors can now penetrate any and all rural regions, no matter how remote, in a way undreamt of even twenty years ago. The long-haul flight, the hire car, the 4-wheel drive vehicle, coupled with the power of television's travel programmes which daily seek new 'undiscovered' areas — all have combined to present a challenge to the continued existence of 'traditional' rurality. Sustainable management systems, defining carrying capacities and regulating access, may be a solution to this type of issue.

Secondly there are the powerful tensions which now exist between the forces of rural development seeking to reverse rural decline, and the forces for conservation, which may seek to fossilise the countryside. Those divergent forces were hinted at in the opening paragraphs of this paper (and see Shaw & Williams, 1994). A sustainable approach may be capable of helping to reconcile the tensions involved.

Thirdly there is the problem of sustaining the rural economy. While there are many problems in developing and managing rural tourism it does offer one positive way of introducing new investment, employment and enterprise into the countryside. A sustainable approach to tourism seeks to avoid an unbalanced approach to economic growth by using tourism as a tool for broader economic progress, actively seeking alternatives to tourism. It also tries to involve local business and communities in ownership, decision-making and benefits (see Frederick, 1992).

There is a final fundamental point: it is the very heart of this paper's *raison d'être*. If rural tourism is special, if rurality in its many manifestations is a unique selling point, then great care must be taken to maintain rurality. Only a sustainable approach can begin to guarantee that condition. That is why it is worth spending time seeking a definition of rural tourism. That is why it is important to realise that rurality varies as a concept from region to region and through time. Ironically, there is a commercial case as well as a case for the conservationist here: both camps need sustainability for long-term success.

References

ADAS (1992) *South Shropshire Pilot Study: Farm Tourism Survey Technical Report.* Wolverhampton: ADAS.

Berry, B.J.L. (1980) *Comparative Urbanization: Divergent Paths in the Twentieth Century.* New York: St. Martins Press.

Best, R.H. and Rogers A.W. (1973) *The Urban Countryside.* London: Faber.

Bouquet, M. and Winter, M. (eds) (1987) *Who From Their Labours Rest?: Conflict and Practice in Rural Tourism.* London: Gower.

Bramwell, B. (ed.) (1990) *Shades of Green: Working Towards Green Tourism in the Countryside* (Conference Proceedings). London: English Tourist Board.

Bramwell, B., and Lane, B. (1993) Sustainable tourism: An evolving global approach. *Journal of Sustainable Tourism* 1 (1), 1–5.

Bryan, B. (1991) Ecotourism on family farms and ranches in the American West. In T. Whelan (ed.) *Nature Tourism: Managing for the Environment* (pp. 75–85). Washington, DC: Island Press.

Carter, H. (1981) *The Study of Urban Geography.* London: Edward Arnold.

CPRE (1994) *Leisure Landscapes: Leisure, Culture and the English Countryside: Challenges and Conflicts.* London: Council for the Protection of Rural England.

Denman, J. and Denman, R.(1990) *A Study of Farm Tourism in the West Country.* Exeter: West Country Tourist Board.

Edgell, D.L. (1993 *World Tourism at the Millenium.* Washington, DC: USTTA.

English Tourist Board (1991) *Tourism and the Environment: Maintaining the Balance.* London: English Tourist Board.

Feifer, M. (1985) *Going Places: The Ways of the Tourist from Imperial Rome to the Present Day.* London: Macmillan.

Flinn, W.L. (1982) Communities and their relationships to agrarian values. In W.P. Brown and P.F. Hadwinger (eds) *Rural Policy Problems* (pp. 19–32). Lexington: D.C. Heath.

Frankenberg, R. (1966) *Communities in Britain.* London: Penguin.

Frederick, M. (1992) *Tourism as a Rural Economic Development Tool: An Exploration of the Literature.* Washington, DC: USDA.

Greffe, X. (1992) Rural tourism, economic development and employment. In *Business and Jobs in the Rural World* (pp. 121–47). Paris: OECD.

Hall, C.M. (1991) *Introduction to Tourism in Australia: Impacts, Planning and Development.* Sydney: Longman Cheshire.

HMSO (1951) *Report of the Commission of Enquiry into Crofting Conditions* (The Taylor Commission). Cmnd. 9091. London: HMSO.

Keane, M.S. and Quinn, J. (1990) *Rural Development and Rural Tourism.* Galway: University College Galway.

Keller, P. (1990) Tourism policy and rural development (unpublished paper for OECD Tourism Committee). Paris: OECD.

Knudson, D.M. (1984) *Outdoor Recreation.* New York: Macmillan.

Krippendorf, J. (1987) *The Holiday Makers: Understanding the Impact of Leisure and Travel.* Oxford: Heinemann.

Lane, B. (1990) Will rural tourism succeed? In T. Hart, S. Hardy and T. Shaw (eds) *The Role of Tourism in the Urban and Regional Economy* (pp.34–39). London: Regional Studies Association.

— (1993) *Rural Tourism: A Bibliography for the OECD.* Paris: OECD.

Murdoch, J. (1993) Review of the economic impact of holiday villages for the UK Rural Development Commission. *Journal of Sustainable Tourism* 1 (2), 145.

OECD (1992) *Business and Jobs in the Rural World.* Paris: OECD.

— (1993) *What Future for our Countryside: A Rural Development Policy.* Paris: OECD.

Pacione, M. (1984) *Rural Geography.* London: Harper & Rowe.

Plog, S.C. (1991) *Leisure Travel: Making it a Growth Market Again.* New York: John Wiley & Sons.

Robinson, G.M. (1990) *Conflict and Change in the Countryside.* London: Belhaven.

Runte, A. (1990) *Trains of Discovery.* Niwot: Colorado, Rinehart Press.

Shaw, G. and Williams, A.M. (1994) *Critical Issues in Tourism: A Geographical Perspective.* Oxford, Blackwell.

Studienkreis fuer Tourismus (ed.) (1987) *Urlaubsreisen 1986.* Starnberg: Studienkreis fuer Tourismus.

United Nations Demographic Yearbooks Time Series. New York: United Nations.

Weiler, B. and Hall, C.M. (eds) (1992) *Special Interest Tourism.* London: Belhaven/Halsted Press.

Williams A.M. and Shaw, G. (1988) *Tourism and Economic Development: Western European Experiences.* London: Belhaven.

World Tourism Organization. *Time Series Data on World Arrivals and Receipts.* Madrid: WTO.

Is Rural Tourism a Lever for Economic and Social Development?

Xavier Greffe

Université de Paris I Pantheon-Sorbonne, 12 Place du Panthéon, 75005, Paris

The development of tourist activities in rural areas has accelerated in recent years, and there is nothing to suggest it will not continue. This expectation of growth is underpinned by a number of factors that can scarcely be concerted. But the problem is now to make rural tourism a possible engine for development, for many reasons. As with any export activity, tourism is a source of income and employment. Because it requires upstream goods and services, tourism serves as a lever for a whole series of activities. The vital question is then how can the services rendered and the products offered be made as varied as possible, so as to be able to attract an increasing number of tourists and maximise the amount they spend locally. Some rural tourism policies must then be implemented: protecting and providing amenities for potential areas of attraction; aiding the renovation and extension of supply; ensuring solvable demand; training people in rural development.

The development of tourist activities in rural areas has accelerated in recent years, but it is by no means a new phenomenon. As early as the end of the nineteenth century in Switzerland, farmers were deriving far from negligible resources from activities connected with accommodating town-dwellers on their farms. But at a time when agricultural activities are able to provide a living only for a diminishing proportion of the rural population, there is thus good reason to consider whether tourist activities offer new opportunities for the economic exploitation of local resources, job creation and enhanced purchasing power.

The development of tourism and the leisure activities that go with it is without a doubt one of the strongest and most obvious trends of the latter part of this century, and there is nothing to suggest it will not continue. This increasing trend deals with the number of tourists and with the lengthening of their stay (Conseil Général de la Vienne, 1992).

This expectation of growth is not only based on the results of empirical surveys, but also appears to be underpinned by a number of factors that can scarcely be contested: new trends in the organisation of working time on an annual basis (OECD Europe), which allows the holiday periods to be split and the use of more economic options; gradual elimination of Saturday working (Japan), which makes it possible to have more short breaks in areas that must not be too far from home, and which thus correspond either to rural or to urban tourism; enhanced productivity in the transport field, making it possible to dampen the effect of energy price increases; longer life expectation and higher incomes for old people, thanks to collective pension schemes, opening up large markets for tourist and leisure activities; and diversification and enrichment of the very concept of tourism, making it possible both to attract new categories of users and where necessary to revive the interest of 'old' consumers.

What is Rural Tourism?

Demand-side definition

This initial WTO definition gives rise to considerable variations, however, the criterion of distance being in practice considered more important than the duration of the visit: it is rare to consider a person travelling less than 75 km as a tourist, regardless of the number of days spent in the place concerned. Thus the United States Federal Task Force on Rural Tourism has adopted the following definition of the term 'rural tourism':

> A visit by a person: to any place other than his or her usual work or home environment and that is outside a Standard Metropolitan Statistical Area; when exercising an activity other than one remunerated in the place visited. The purpose of such a visit may include pleasure, business, study, health or religion.' (United States Congress, 1989).

This approach based on the nature of the visitors gives rise to two problems, however: first, through not being standardised it cannot lead to fully comparable relevant statistics; second, it fails to take account of the dynamism of the supply side which, in a period of reorganisation of the rural tourism 'production chain' such as we are experiencing today, is of vital importance for the prospects of the activity. We shall therefore look more closely at this second, supply-side, approach.

Supply-side definitions

On the supply side there are three possible ways of understanding the term 'rural tourism'. In the first place, rural tourism is understood as staying with a local resident in a rural area. Thus in Portugal, rural tourism seems to be associated with lodging local people in a cottage or on a farm, which tends to reduce the content of rural tourism policy to the problem of extending the accommodation capacity (EEC, 1987: 67).

This first approach leads to the idea that rural tourism has something 'poor' about it: it is the poor man's tourism, for people lacking the means either to go to the major 'tourist' areas or even to rent self-catering accommodation in the more economically accessible areas. This image of a poor man's tourism doubtless does not apply everywhere, however: it is quite marked in France, where co-operative and mutualist organisations have played a big developmental role, but this is not at all the case in Germany or Austria, where better-off social groups have always indulged in rural tourism.

A second approach to rural tourism therefore associates a particular form of accommodation with the possibility of undertaking specific activities, an essential step towards understanding the contemporary concept and content of tourism and a means of breaking away from the attitude that rural tourism is a poor man's holiday. On this activities side, the possibilities are increasingly numerous: cycle tours, fishing, riding, discovering farm life, getting fit, etc. (Grolleau & Ramus, 1986: 14).

Understood in this way, local tourism thus gives rise to the most varied scenarios, and what the approach gains in terms of highlighting the services offered, it loses in terms of precision. Accommodation extends from staying with a family on a farm or elsewhere, through a self-catering cottage to a small hotel. The space for this type of tourism may be farm or forest, water or air, and the tourist's purpose may be pleasure, health or even education (for example, the so-called 'green classes' where in some European countries certain classes from an urban school spend a period during term, usually in mountain regions).

A third approach has emerged over the last few years. This is a type of holiday based in principle on being housed on a farm and organised around activities connected with the farm. There is thus a double precision: the place of stay is clearly defined, even though there may remain a certain latitude as to its specific nature: self-catering cottage, room in the farmhouse, rented accommodation, etc. The activities are clearly defined, as here they are those that may traditionally be met with on a farm, including the discovery of its products and the local area. Thus activities such as cures at hot springs are no longer covered by the rural tourism umbrella, an unjustifiable assimilation in the eyes of some, perfectly legitimate in the eyes of others. This approach is therefore encouraged all the more in that it appears from the standpoint of the farmers to be a way of compensating for the drying up of certain traditional sources of income, and from the standpoint of the authorities to be a way of relieving the overcrowding of the traditional tourist sites and the disamenity this causes. It may be asked, however, whether this type of approach is not likely to very soon come up against fundamental limits, especially at a time when the concepts of space and rural activity are undergoing far-reaching changes: today farmers are developing or imagining activities which have little to do with the 'farm' in the traditional sense and which are tending to establish a bridge between the farm holiday and tourist activities located in other spaces, such as golf, riding or fishing.

Actually, it would be wrong to make rural tourism the systematic consequence of the difficulties encountered by agricultural policies, and even more so to try to make it the answer to their problems.

Under these circumstances, rural tourism can offer a partial solution, acting in three ways:

- It enables other possibilities for rural activities to be defined. As a direct consequence of what has just been said above, in all the countries studied it is considered that the 'rural tourism' product should be aimed at solvable demand (market-oriented) and managed by private structures resorting to public aid as little as possible except perhaps for the establishment and launching phase.
- It makes it possible to create outlets for traditional products some of which are in over-supply (farm produce) and others tending to disappear (handicraft products).
- It also makes it possible to create new sources of income and to transform casual jobs for women, sometimes even creating permanent jobs, though it is rare for new jobs of the traditional type to be created.

From Rural Tourism to Local Development

Rural tourism is considered to be a possible engine for development. There are a number of reasons for adopting this attitude.

- As with any 'export' activity that is organised and developed, tourism is of course a source of employment and income.
- Because it requires upstream goods and services — agricultural products food and fishing industry products, clothing and textiles, building and public works, transport, insurance, etc. — tourism serves as a lever for a whole series of activities, some of which are already present or even inherent in the rural context. In this regard it must never be forgotten that rural tourism forges a whole chain of activities and is not simply one industrial stage among others.
- Rural tourism exploits and 'capitalises' the countryside or natural phenomena to which no economic value is attributed. It amounts to a kind of exploitation of a deposit as in the mining or oil industry.
- Tourism promotes the demand for craft work and labour-intensive products which tend to be frequently met in the rural environment.
- To the extent that this rural tourism attracts tourists from outside the economic space in question, it stimulates the injection of exogenous flows of expenditure and brings about traditional multiplier effects.

Looked at in this light, rural tourism goes far beyond the stage of simply complementing traditional agricultural activity, the role most often attributed to it. It becomes a perfect example of the 'Travel — Tourism — Leisure' chain and leads to forms of organisation that are far from remaining at cottage industry level. The logical partners tend to be transport undertakings, travel agents, financial services, agricultural or rural organisations and local authorities, so that the farmer finds himself 'endowed' with many partners. There then arise inevitable problems connected with the control of the system and the sharing of profits. Rural tourism may be the main instigator of all this activity, but it is no longer by any means sure of being the main beneficiary.

There is by no means universal agreement about the beneficial effects associated with rural tourism however, and there are some powerful counter arguments:

- The development of rural tourism is often experienced as an aggression against an environment and way of life, a disturbing factor with no real compensations. The argument has some truth in it, but nevertheless needs some modification in view of the disappearance of many of the traditional activities of the way of life its proponents want to protect: the fact that certain rural habitats are already being destroyed, rural tourism or no rural tourism.
- Organising tourism costs a lot of money, often more than is brings in. Here again, there is some truth in the argument, but it all depends on whether two things can be achieved:
  - (i) stimulation of autonomous expenditure above and beyond the traditional hotel and catering trade expenditure where the multiplier effect is fairly limited;
  - (ii) a fair share for the local rural area of the profits created by the tourist chain.
- We now arrive at the third argument which is in fact the vital question: how can the content of tourism in rural areas be 'enriched', i.e. how can the

services rendered and the products offered be made as varied as possible, so as to be able to attract an increasing number of tourists and maximise the amount they spend locally. It is not easy to find an answer to this question as it requires investment in human resources, equipment and even imagination, and experience shows that these are not always co-ordinated in the most efficient fashion.

Demand for Rural Tourism

According to the findings of surveys carried out in several countries, the profile of the 'holiday-maker' in rural areas is undergoing profound change. Thus a study carried out in Germany by the Studienkreiser fuer Tourismus (Grolleau, 1988: 72) shows that it is the middle and upper socio-professional categories who dominate the sample of those interested in rural tourism, and 53% of these holiday-makers have some higher education, as against 41% for the holiday-makers as a whole. What is more, the reasons for choosing rural tourism, in decreasing order, are: contact with nature, desire for family holidays, and in third place only comes the desire for an inexpensive holiday. Similar results are found in a number of Northern European countries.

However, a similar pattern is now found in the Mediterranean countries too, which is more striking in that the model that dominated there not so long ago was 'rural tourism is for the less well-off'. Thus an Italian survey (*Agriturist*, 1985: 27), shows that the people who enjoy rural tourism are between 30 and 50 years old and 79% have completed middle level or higher education (which is far above the average educational level of holiday-makers) and the desire for peace and quiet ranks equally with the desire for a fairly economical holiday (a motivation that thus appears less marginal than in Germany). In France too, surveys confirm the emergence of this new profile: the demand for rural tourism is no longer from the less advantaged categories for whom lodging with a family is the only way of escaping from the town, or even from new town dwellers who want to return to spend some time in their native villages: the tourists are generally people with no prior link with the chosen rural area, and who subsequently hope to return.

It is thus possible to put forward the idea that the increase in tourist expenditure should benefit rural tourism more than other forms (for example seaside holidays), provided, of course, that the supply structures are able to attract and retain these new potential clients, which implies investment in the provision of services.

The appearance of this new demand profile for rural tourism has been confirmed in recent years by market surveys carried out using commercial management methods. Thus a survey carried out in Norway for the local authorities in Telemark was commissioned to determine what communications media should be used for publicising rural tourism in the area, this appearing to be one of the possible strong points of the region, both because of its countryside and environment and because it is relatively close to the major Nordic towns.

This analysis, based on a correspondence break-down, made it possible to identify two structural axes: 'materialist–idealist' on the one hand and 'traditional–modern' on the other. Four broad categories of tourist were then identified as shown in Figure 1:

- in the upper right quadrant people looking for theme holidays;
- in the upper left quadrant people looking for 'sex, sand and sun';
- in the lower left quadrant people looking for comfortable holidays with everything booked in advance;
- in the lower right quadrant people looking for family holidays.

MODERN

SEX, SAND AND SUN **THEME HOLIDAYS**

MATERIALIST **IDEALIST**

HOTEL HOLIDAYS **FAMILY HOLIDAYS**

TRADITIONAL

Figure 1 Major categories of tourism

This analysis makes it possible to see that the people interested in rural tourism are situated in the two right hand quadrants, and mainly in the upper quadrant. Knowing the distribution of the socio-professional categories between the four quadrants, it is possible to identify the people with a potential demand for rural tourism; relatively young or middle-aged with a good educational level and family responsibilities. The Telemark authorities therefore designed a publicity campaign specifically targeted to attract this type of clientele, notably through using the press they read.

The Supply of Rural Tourism

The tourism supply in rural areas remains very much marked by the types of accommodation that predominate and the quality of the holiday is fundamentally affected by this factor, but in recent years the diversification of services has made it possible to realise this approach in terms of type of accommodation.

A supply characterised by the accommodation situation

The initial specific categorisation of rural tourism in terms of accommodation is found every time one wants to evaluate the tourist capacity available and the problems encountered. Though there is no universally accepted typology here, the available accommodation may be broken down as follows:

- hotels, generally of a medium level of comfort, though hotels play a marginal role in many cases;

- camp sites;
- family holiday homes, generally with some social affiliation;
- holiday villages and self-catering accommodation, generally individual houses, grouped or scattered, with or without restaurant facilities and leisure activities;
- seasonal renting of furnished accommodation;
- board and lodging on farms or with other local residents;
- secondary residences.

The list of different types of accommodation may be extended in some cases, or may be combined together as with the growing number of 'staging post' arrangements. The distribution of the different categories of accommodation varies considerably from one country to another, as is shown by the comparison in the following tables between Norway (Table 1), a country where rural tourism has long been practised, and France (Table 2), where its significance has changed fairly considerably in recent years. In fact it can be seen that the importance of secondary residences is historically determinant, as it suffices to reverse the respective shares of the 'commercial' and 'non-commercial' sectors.

The EEC data on the other hand, constructed on the basis of different categories, bring out certain features common to the various member states, but these

Table 1 Breakdown of types of accommodation in Norway

Commercial sector of which:	20%
Hotels	5%
Camping and cabins	3%
Rented chalets	4%
Private sector of which :	80%
Personal chalets	32%
Accommodation with friends or family	40%

Source: Ministry of Industry, Oslo, 1989.

Table 2 Breakdown of types of accommodation in France

Commercial sector	43.3 %
Hotels	5 %
Camping	21.7 %
Rented accommodation	16.6 %
Private sector	49.5 %
Main residence of friends or family	26.2 %
Secondary residence of friends or family	9.5 %
Secondary residence	13.8 %
Other	7.2 %

Source: Grolleau, 1986: 73

have to be treated with caution because the items concerning rural tourism are interpolated between general statistics on tourist accommodation conditions (EEC, 1987: 84). Here hotels and boarding houses remain in first place with almost 30% of stays, followed by staying with friends (21%). These are followed by camping with almost 16% and rented accommodation (16%). Secondary residences account for only 10% of stays.

Just as difficult as this static interpretation, is that of trying to determine dynamic tendencies valid for all countries, as their economic trends and their legal structures are far from corresponding with one another.

- The hotel trade is progressing little and has difficulty in adapting to the standards of comfort demanded by the new clients of rural tourism. Thus the problem encountered here is more one of high cost than of comfort or quality.
- Conversely, there is considerable expansion in self-catering cottages, 'holiday villages', etc. Non-existent in 1960 in a country like France, their capacity had already reached 10, 000 beds in 1970 and 250, 000 in 1985.
- Boarding with local residents has been growing rapidly since the beginning of the 1980s, but it should be noted that the practice has long been institutionalised in the German-speaking and Nordic countries (in Austria, almost 26% of farms have been receiving holiday-makers for almost 100 years now).
- Apart from staying with family or friends, camping remains fairly generally the principal form of accommodation.

The statistical problems encountered during the static study of types of accommodation have no reason to disappear when the dynamic aspect is being considered, so once again it is difficult to make international comparisons. It should be noted however that in the case of a country like France the broad trends outlined are on the whole verified, as shown in Table 3.

It will be noted in particular that the type of accommodation we have referred to as belonging to the social economy has grown very substantially (up 65% in the case of non-commercial holiday villages and 74% in the case of family guest homes). Also to be noted is the fall in non-classified hotel accommodation, a reflection of its poor quality.

Table 3 Trends in types of accommodation 1971–1991

Type of accommodation	*Variation in the number of beds*
Approved tourist hotels	+ 15 %
Non-classified hotels	- 25 %
Camp sites	+ 70 %
Secondary residences	+ 34 %
Non commercial holiday villages	+ 65 %
Rented cottages, bed and breakfast	+ 140 %
Family guest homes	+ 90 %
Youth hostels	+ 65 %

An increasingly varied supply of services

An increasingly varied range of services is being grafted on to these alternative forms of accommodation, and this is indeed the way to interest potential clients in rural tourism. The last few years have seen the multiplication of very varied possibilities from rambling to air tours (including by hot-air balloon), the range of activities including fishing holidays, pony trekking, introduction to traditional crafts, the general discovery of wild life and farm animals, and specialised attractions based on specific types of animals such as sheep or horses or bird-life.

This development has very important consequences from the standpoint of the organisation of supply. Such activities no longer involve a fixed stay, but imply daily moves and constant coming and going between different forms of accommodation or between different 'partners' in a single form of accommodation. *The supply therefore needs to be organised in networks*, the services provided now becoming more of an attraction than the actual accommodation conditions. This implies the sign posting of circuits or trails used for a specific purpose.

The supply needs to be of an increasingly high standard: it no longer suffices for a hotel to supply fishing rods or a place where they can be kept. What is needed now is a high quality of service in terms of the hire and maintenance of equipment and information, logistic support, etc. There is nothing to stop the providers of accommodation from supplying such services, but the connection is by no means automatic, even though it should follow naturally if it is intended to offer such services at an acceptable price and ensure that the greater part of the tourists' expenditure is retained for the benefit of the local population. In addition, this implies a high level of training in certain fields.

From economies of scale to economies of scope

Setting up new rural tourism systems therefore implies organisational difficulties. But it also opens up a new opportunity, that of economies of scope, whereas the law of traditional tourism had become that of economies of scale. It should be noted in fact that the main consequences of the concentrations realised in seaside and ski resorts in order to achieve economies of scale have been escalating costs, environmental damage that in certain cases appears irreversible and the disaffection of increasing numbers of tourists, hence problems of amortisation and the setting of exorbitant prices.

Other economic strategies become possible with rural tourism, which is in fact increasingly turning out to be a joint provision of several services and products from a relatively well-identified location (or itinerary). This opens up the possibility of deriving substantial revenue no longer through maximising the number of tourists but rather through maximising the opportunities for them to spend money, which they may well do in view of the fact that some of their items of expenditure are in any event very reasonable as compared with other forms of tourism. Specialists are broadly agreed that it is interesting to note that the new tourists in rural areas are often those who have become disenchanted with more traditional forms of tourism which have become overcrowded, so that any brake on such a strategy would not be a monetary brake in this case.

Rural Tourism Policies

The role of the authorities becomes clearer in the light of the above analyses of demand and supply: the supply conditions need to be renovated and extended in order to attract and hold on to a strong potential demand. Their role thus consists of several strategies:

- protect potentially attractive areas and make infrastructure investments to their advantage;
- modernise the supply structures and production of tourist services in rural areas;
- promote these structures in the eyes of the potential users, as is already done by certain private producers' organisations, but not systematically at present;
- help make potential demand solvable: certain categories of rural tourist do not have the means to pay for costly forms of tourism;
- train rural people for their new activities.

On top of these strategies, however, it is necessary for the authorities to consider other problems, starting with the question of knowing whether or not to concentrate their intervention on clearly defined geographical areas or on the contrary to spread their action over the whole of the territory concerned.

Protecting and providing amenities for potential areas of attraction

The first stage in an authority's strategy for rural tourism consists of protecting potentially attractive areas and ensuring good access to them. *A priori* the action here is to protect the environment and milieux concerned, and generally speaking there is no lack of agencies and institutions working in this direction today, even if this may not be their primary motivation. This makes the instruments used all the more varied: inventories, technical assistance, legal measures and regulations, financial aid and incentives, support with the organisation of particular events, support for traditional cultural activities and crafts, direct financing of rehabilitation operations, etc. Public utilities may be questioned at this first stage: e.g. it may be required to suppress electric wires, and put them underground (as already realised in Poitou-Charentes at Nouaillé-Maupertuis).

These actions and interventions may come into conflict with other concerns, notably those of farmers or landowners who do not want to see the development of rural tourism. This results in a difficult task for the authorities: determining the spaces within a given geographical area that can be appropriated for tourism and those that cannot.

Norway has gone quite a long way in this direction, drawing up land use maps for rural areas, then discussing them with all the local partners. The process begins on the basis of surveys of past and foreseeable practice, in order to distinguish between zones where there is likely to be conflict and those where there is not. Attempts are then made to find mechanisms that may reduce the number of zones of conflict: establishment of reserves, creation of new tourist routes. Lastly the zones where the conflicts of interest cannot be reconciled are identified and submitted to the judgement of the local population concerned. Once the zoning is completed, a transport plan, a cultural programme and an economic development programme are added to this initial map.

The establishment of transport programmes in fact appears essential. As has already been pointed out it is necessary to have good access to a rural area and to be able to travel within it. Thus improved communications in rural areas is one of the modifications most frequently encountered; this improvement may lead to the creation of scenic roads, or, somewhat to the contrary, adventure trails.

In addition to transport infrastructure, the needs in terms of communications are very important: not only the integrated organisation of different modes of transport, but also local telephone services, fax, financial communications, etc. The problem here is that the organisation of such infrastructure is never considered a priority, whereas in fact it is part and parcel of the requirements of the new rural tourists whose profile has been identified.

Aiding the renovation and extension of supply

In all the countries where active policies to promote rural tourism are in force, the measure that is always found is financial aid to persuade the owners of accommodation to modernise and extend their premises. Thus in Austria, Italy and Ireland, there have been many schemes of this type, based on the principle that boarding with local residents, notably on farms, does not always provide the level of comfort that tourists expect, and modernisation investment is too expensive, at least for those who cannot be reasonably certain of deriving adequate additional income from such investment. In addition, since the banks hardly fall over themselves to finance such operations, it is left to the authorities to find the solution.

In this context, Belgium has organised systems where 25 to 30% of the cost (subject to a ceiling) is covered by the authorities. Portugal has recently set up a system for financing up to 75% of the cost of modernising an installation recognised as being of public interest. Since 1967, Spain has been offering incentives for investment in accommodation through the system of Casas de Labranza, which led to capacity being tripled in twelve years (EEC, 1987). More recently, Italy too has adopted a scheme extremely favourable to this type of rehabilitation in rural areas (Law of 5th December 1985). In France, the department of the Vienne has adopted a comprehensive system of financial incentives: investments in hotel accommodation is supported by up to 10% of their cost, and investment in farm lodging, family homes, etc. by up to 25% of total cost. The results have been stupendous: the relative share of the best qualified accommodation has increased by 6% for the hotels, 60% for farm lodgings and 55% for camp sites, between 1987 and 1992.

Another type of action by the authorities is aid with management. Thus in the United States, the Small Business Administration organises advisory and training services for small businesses in the tourism sector through mobilising a substantial network of consultants and experts including retired executives as well as academics. This has resulted in many workshops devoted to subjects as varied as retail trade in rural communities, home-based businesses, the promotion of tourist products and managing businesses in the hospitality industry (US Congress, 1989: 33).

Yet another type of action in favour of modernising the supply is now starting to be undertaken in a fairly systematic way: the setting up of information programmes in the rural world concerned with receiving tourists. These programmes generally appear in the context of further occupational training, but are also found further upstream, with certain agricultural colleges integrating basic technical and commercial knowledge into their courses.

Promoting rural tourism

The authorities' third broad strategy is to undertake actions to promote rural tourism. In some countries the suppliers have grouped together for this purpose, but this is by no means the case everywhere, and in many places the authorities intervene, either to bridge gaps in such networks or to consolidate their action. Thus in Belgium it is the Office de Promotion Touristique that assumes this function, private initiative having turned out to be inadequate. In Ireland, the Irish Tourist Board is concerned with the technical side of non-conventional accommodation, complementary leisure activities and the supply of the corresponding information. In France, the best tool is the 'Office du Tourisme Départemental' which offers a lot of informational support: brochure, telematic treatment of enquiries, and participation in international tourism trade fairs such as ITB or Utrecht.

These promotion activities quickly lead on to the problem of transport, for both national and foreign tourists. Generally speaking, the existence of rail and road transport networks concerned with the objectives of rural tourism is a precondition for success. Unfortunately however, good access is not always provided, the seasonal nature of rural tourism often discouraging initiatives taken in this direction. The mobilisation of air and sea transport also appears necessary, and in certain cases turns out to be quite successful (Ireland). This introduces another problem however: when airlines agree to move in this direction, this puts them in a position of being essential intermediaries and they tend to impose a sharing of the benefits that is unfavourable to their partners (as in the Scandinavian countries, for example). This shows that the export of rural tourism programmes implies not only the co-ordination of all the actors concerned, but also the creation of a fair partnership system.

The systematic labelling of the programmes offered is also becoming one of the objectives for the authorities, and this effort is found in countries like Spain and Portugal. In other countries however, the authorities have tried instead to encourage existing private structures to group together to form a unified framework for the supply of information. This is the case in France, where the 'Tourisme en Espace Rural' movement brings together three federations: the National Federation of Rented Cottages ('Gîtes Ruraux'), the National Federation of Inns and Hostelries of France and the Holiday Village Association.

Closely associated with labelling is the effort of certain authorities to provide rural tourism suppliers with guidelines indicating the different conditions to be fulfilled, though without this necessarily leading to a sanction in terms of receiving a classification or not (Grolleau, 1986: 32).

A third action under this strategy is for the authorities to facilitate the intercommunal links necessitated by rural tourism and to enhance the living environment beyond the standard traditionally maintained by private initiative. Thus in the Netherlands the local authorities have organised themselves for the joint management of adjacent areas and to facilitate the creation of the itineraries that often form the base for many touring activities in rural areas (Rekreatie Schappen). In France, as in Ireland, there are many 'floral village' competitions organised to set off the rural heritage, to say nothing of the measures taken with regard to the environment, protection of the countryside, etc.

Ensuring solvable demand

While the new clients for rural tourism seem on average to be better-off than the more traditional ones, it should not be forgotten that an important fact still remains: for many families on middle or low incomes, rural tourism represents one of the few opportunities open to them. This becomes very evident when one considers the importance of the mutualist, associative and co-operative movements in the organisation and production of accommodation capacity. The authorities cannot ignore this segment of the clientele, and should facilitate the solvability of this demand through attracting other sources of additional finance or through decreasing the real cost of the services provided.

The holiday voucher mechanism now adopted in several countries goes in the first direction. It has tax advantages from the standpoint of the employers or trade unions, who can thus increase purchasing power for a specific purpose. In rural areas however, this holiday voucher mechanism comes up against a number of obstacles which need to be taken into account before launching any action in this field:

- the traditional individualism of the rural world means that the concept of the subsidisation of visitors may be badly received by farmers;
- this method of payment exposes people to the risk of tax inspection in conflict with certain habits in rural areas;
- this type of tourism, if heavily concentrated gives the local actors a share of the income that is less favourable to them than that given to the transport and financial enterprises involved in the organisation of the system.

The second direction taken by the authorities is to limit the necessary expenditure by making it play several roles, but at the same time act as a type of pass. Thus the authorities in the Austrian Tyrol decided not to eliminate a tax on holiday stays — the justification of which may be open to question — but to make this tax equivalent to a free insurance policy for all the activities engaged in during the holiday, with it even giving the right to reductions on the price of those activities. (Grolleau & Ramus, 1986: 317, v. 116–120)

Train rural people in rural development

In addition to the overall problem of trying to achieve a net positive variation in the number of jobs, the development of rural tourism raises another: that of the specialisation of activities and the resultant need for specific training. The problem of adjusting human resources is not simply that of stimulating the farmers' capacity to offer hospitality to 'city dwellers', housing and feeding them, which is the somewhat simplistic view often adopted. In view of the transformation of rural tourism and the different motivations of the potential visitors, there is need for activity organising, management skills, etc.

A recent typology shows the gap that exists between what could be considered just a few years ago as adjustment at the margin and what is now becoming a veritable human resources training policy. In addition to jobs in the catering trade, there are activity leaders and tourist guides on the one hand and holiday, circuit and itinerary management jobs on the other. The different qualities required are not readily substitutable or calculable, all the less so as the degree of professionalism required can only increase.

Many countries have risen to the challenge by organising information or awareness-rousing courses, using conventional channels (open universities in the Nordic countries). Other countries, such as Germany and Austria, leave it to local organisations to disseminate the relevant information, while others again incorporate such matters in initial vocational education.

The Norwegian programme for diversifying the skills of agricultural advisors is worthy of mention here. The Ministry of Agriculture has a body of almost 2000 advisors who intervene at the request of farmers to give both technical and commercial support. In view of the expected redeployment of activities, and the decline in a number of them, it has been decided to diversify the competences of these advisors by training them to be able to give support in the field of rural tourism, which means that first they themselves must be trained in these new activities. This programme is expected to take two years, after which there should be a snowballing training effect in favour of all those farmers who intend to become involved in rural tourism. In parallel with this programme, Norway is developing local tourism training programmes in colleges and universities, with the possibility of choosing one of three specialisations: economics of local tourism, transport and equipment planning, hotel and supply management.

In Wales too a specific programme has been set up. The development agencies, starting with the Wales Tourist Board, have introduced a programme called 'Getting into Rural Tourism', intended to train people in the management of farms receiving paying guests. Further upstream, conventional agricultural colleges are now including an introduction to rural tourism practice in their courses (LEDIS, 1989b: 3).

Should rural tourism policies be centralised or decentralised?

Many states have defined and implemented policies for rural tourism in relationship with the previous objectives. But we have to ask ourselves about the relevant level for organising such policies. More precisely, should these policies be organised at central level, intermediate or regional level or village level?

The interest in implementing rural tourism policies at the intermediate level is very important. Two reasons may be given:

- It has been underlined that rural tourism activities must be developed in a network perspective in order to increase the supply of touristic products. This implies cooperation beyond the village level, but relevant to an intermediate level.
- In rural tourism you have to choose what are your target groups and how to attract them in your area. This cannot be realised without an intermediate level: a central level is unable to specify this policy in relationship with a

specific environment; and a village level has neither views nor ways to organise such an attraction.

The example of the Vienne Departement in West Central France illustrates this requirement. This Department enjoys the existence of a very important leisure park, the Futuroscope which was visited by 1.2 million people in 1992 and which expected 2 million visitors in 1993. On the other side, the Departmental of Vienne includes some beautiful rural areas where the main products or activities of rural tourism may be developed.

The Department has organised itself to make rural areas benefit from the Futuroscope touristic agglomeration. All the ways and means feasible have been mobilised: information networks at the core of the Futuroscope activities; mobilisation of investments to provide better accommodation; specific training at rural colleges; investment in environmental improvements and protection, etc.

The results already appear beneficial and the inflow of rural tourists may be close to 50,000 (20,000 foreigners and 30,000 French citizens), which would be a very good result. This underlines the relevance of rural touristic policies at the intermediate level.

The Effect of Local Tourism in Terms of Employment

There are very few studies on the effects of local tourism on employment, there being three factors that clearly limit the possible content of such a debate:

- The number of variables acting on the determinants of employment is as great at local level as at the macro-economic level, but the identification and accounting systems are less effective in view of the 'open' nature of local spaces;
- It would seem that as regards accommodation and catering expenditure, the budget is used more to maintain existing employment, even by means of redeployment, than to create new jobs.
- The effect of expenditure other than on accommodation and catering, thought to have a significant multiplier effect, is little known at present: the employment functions of rural tourism are difficult to analyse because of the lack of detailed surveys. In certain cases findings are available however: thus a SOGEL study considers that a golf course creates three times a many jobs as a farm on the same land, but in other cases estimation is more difficult to analyse because of the lack of detailed surveys. How many jobs does a fishing itinerary create?

Maintenance and transformation of existing employment

In the arguments associated with public policies for local development two themes constantly recur:

- expenditure on ancillary activities turns out to be most relevant to induced job creation;
- the value of rural tourism lies above all in the complement it brings to agricultural incomes, which tends to confirm the thesis of the defence of existing jobs rather than the creation of new ones.

In France, a study by the Villages Auvergne Limousin association, which manages 28 villages with some 7,500 beds, illustrates the validity of these arguments and its findings are close to those of the Norwegian study mentioned above in the analysis of expenditure multipliers. Over a season (1984–85), the data collected showed that the ratio between additional expenditure and accommodation expenditure ranged from 90.7% in holiday residences for old people and 91% in boarding houses, to263% in 'normal' accommodation and 323% in rented cottages.

Two important conclusions emerge from this:

(i) In all cases, the additional expenditure represents a sum that is far from being marginal, and it has to be considered that even at the minimum it is at least as great as the reference expenditure, to which everyone alludes. This clearly shows that one of the objectives of rural tourism should be not only to maximise accommodation capacities but to offer opportunities for additional tourist expenditure through providing as broad a range of services as possible.

(ii) No doubt the ratio reported varies logically enough with the accommodation cost: if this is high, other expenditure being lower and vice-versa. But the final differences in additional expenditure are considerably greater than the initial differences between accommodation costs: 358% as against 69%. The interpretation according to which tourists make a trade off between what they are prepared to spend on accommodation and what they prefer to devote to purchases and pastimes is thus inadequate. The interpretation according to which old people spend less than others on entertainment is certainly more plausible, but the sample here is of limited significance and the phenomenon is in any case well-known, and not only from the case of rural tourism.

This study subsequently went on to make a general estimation of the work-equivalent of all this expenditure. This amounted to 2532 man-months or the equivalent of 196 full-time jobs, but in fact four-fifths of these jobs were filled by seasonal workers, a considerable proportion but not untypical of activities in agricultural areas. Instead of seeing this as proof of the poor quality of the jobs concerned, it is better to look at it as showing that they are adapted to the rural context, though with one major difference: while many of the non-agricultural activities in rural areas tend to appear during the winter, tourism-related occupations tend to be concentrated on the summer. Competition between activities is not very great, and may turn out to be a positive factor where farming activities are in decline.

The development of agritourism in Greece has also helped to demonstrate rural tourism's role in maintaining jobs and incomes. What is more, it is interesting to note here that these objectives, seen at the outset as being the positive effects of agritourism, were achieved through the creation of co-operatives. Thus on the island of Lesbos, the creation by women of a co-operative to develop agritourism in August 1983, had led five years later to the creation of 30 seasonal jobs, 5 male and 25 female. The male jobs were essentially in catering and the female in activities connected with accommodation, visits and excursions (LEDIS,

1989a: 1). In Prespes, in the north-east of Greece, the creation of another co-operative with a capacity of 30 tourists led to the creation of 20 seasonal jobs. There are now seven co-operatives of this type, or a total of almost 150 jobs. These initiatives tend to be in the consolidation phase at present, but others are being created with the support of the authorities, their launching always causing problems from the financial standpoint (LEDIS, 1989a: 2).

Two Japanese Ministry of Transport studies also bear witness to local tourism's ability to maintain or even create employment. The first study concerns the Api area, situated in the Iwate prefecture in the north of Japan. It is interesting to see what the effect of a major transport innovation can be on a hitherto unexploited rural area. In this case, the introduction of a very high speed train cut the journey time from Tokyo by more than half, to only two and a half hours for the 515 kilometres. By 1986, four years after the introduction of this service, the number of wage-earners working in this zone in jobs specifically connected with rural tourism had increased from 22 to 312, the tax revenue had risen from 12 to 124 million yen, the net income of the hotels from 5 to 796 million yen and in catering establishments, from 12 to 292 million yen, all this with no major innovation other than the new train and hence the greatly reduced journey time, obviously a major factor in a country where the amount of free time is very limited.

The second study was concerned with the Toman area on the island of Hokkaido. This had been converted to offer winter sports in winter and such activities as golf, tennis and archery in summer. Over a period of five years (1980–85), the number of jobs in this small area increased from 400 to 1000 the number of permanent inhabitants from 1601 to 2097 and the tax revenue from 8.3 to 285 million yen.

While the first experience showed the importance of transport innovation, the second highlights the value of diversifying activities and thus maximising the capacity utilisation of the hospitality facilities, which in this example can be used all year round.

From seasonal to permanent employment

Because of its concentration in time, rural tourism generally appears as a factor creating jobs that are episodic, rather than permanent, but still for a longer period than certain traditional agricultural activities. Thus a survey in the Poitou-Charentes region in France showed that in rural areas two types of employment are tending to disappear: seasonal jobs of less than 39 days a year and virtually permanent jobs of over 245 days a year. The number of full-time jobs fell by 34% between 1979 and 1985. On the other hand, the supply of jobs for a duration of between 39 and 245 days a year remained stable, at least over the period covered by the survey, i.e. the first half of the 1980s. These jobs appear to be concentrated mainly in the field of rural tourism, silviculture, clearing undergrowth and pig production. The fact remains that through creating essentially part-time and/or seasonal jobs, rural tourism can end up merely competing with other activities without any net gain to the community concerned, and does not necessarily lead to social progress.

It is thus in the general interest for employers to group together in such a way that by combining their activities they can create one stable full-time job where there were previously several seasonal jobs. It should be pointed out that the demand here comes from the employers rather than the workers, even though the latter can but benefit. The reason for these 'employer groupings' is not, as may be thought at first sight, fiscal: there is no exemption from fiscal charges for the creation of the first job, and the grouping enjoys no special advantages since it is subject to corporation tax while it is in fact a non-profit organisation.

Some Final Remarks

The different questions that have been raised above show that rural tourism is far from being a panacea for all the difficulties now being met with in sparsely populated areas located far from the big conurbations and where the agricultural activities that formerly constituted the mainspring of development are facing a bleak future. In general terms it can be said that the capacity of rural tourism to contribute to the resolution of the problems faced by rural areas depends on the solution of a certain number of problems that may be summarised as follows:

- The quality of the accommodation and the variety of services offered constitute powerful levers for the development of rural tourism. If full advantage is to be derived, the approach cannot be other than 'professional' and this implies a certain minimum of training and preparation.
- Since rural tourism forms part of a whole, which may be generically designated 'tourism — transport — leisure', the local actors will benefit from the incomes thus generated only if these incomes are not exclusively controlled by the upstream stages of the system (transport, publicity or financing services).
- The common, but nowadays often self-defeating, search for economies of scale in tourism should give way to a search for economies of scope in order to try to maximise tourist expenditure rather than tourist numbers. In the domain of rural tourism, this search is deeply relevant.
- Obtaining results in terms of jobs will depend on the ability of the supply side of rural tourism to go beyond merely providing accommodation and assume a service supply content; in any event, the effects in terms of jobs will consist in the first place of maintaining existing jobs and thus stemming the rural exodus.

Bibliography

Agriturist (1985) Lo sviluppo dell'agriturismo. Roma.

— (1987) Esperienze e prospettive dell'agriturismo. Roma.

Arzeni, S (1989) Turismo Internazionale: evoluzione e tendenze, Scenari, Juglio-Agosto.

— (1989) International tourism: Past trends and prospects, IIEC Discussion Paper, Fall.

Ascher, F. (1984) *Tourism: Transnational Companies and Cultural Identities.* Paris: UNESCO.

Bote, V (1984) *Importancia de la Demanda Turistica en el Espacio Rural en España.* Madrid: Consejo superior de investigaciones scientificas.

Business Week (1989) Greyhound is bringing travellers down to earth again. 19th June 1989.

— (1987) Theme Parks: This Slugfest is no fantasy, Disney and MCA are going head to head in the $4 billion business. 23rd March 1987.

Cannata, G. (1987) *La domanda de agriturismo nelle campagne italiane.* Anagritur.

Conseil Economique et Social (1984) Les aspects économiques de l'industrie du tourisme. *Journal Officiel*, Paris.

Conseil Général de la Vienne (1992) *Rapports sur les activités touristiques*. Poitiers.

Dernoi, L.A. (1983) *Situation actuelle du tourisme rural en Europe.* United Nations.

EEC (1986) *Europeans and Holidays.* Brussels: General Directorate for Transport, June.

— (1987) *Rural Tourism in the Twelve Member States of the European Economic Community* Brussels: General Directorate for Transport, November.

Gamble, P.R. (1988) Tourism technology: Developing effective computer systems. *Tourism Management*, December.

Grilleau, H (1986) *Analyse du tourisme rurale.* Direction du Tourisme, INSEE.

Grolleau, H. and Ramus, A. (1986) Espace rural, espace touristique: le tourisme à la campagne et les conditions de son développement en France. *La Documentation Française*. Paris.

Hawkins D.E., Shager, E.L. and Rovelstad, J.M, (eds) (1980) *Tourism Marketing and Management Issues.* George Washington University, Washington DC.

Hol Prosjektet (1989) Reiselivmaeringen og den lokale o008/konomi, BRR, Oslo.

Lebaube, A. (1989) Syndicalistes et organisations professionnelles jouent la carte du développement rural. *Le Monde*, 17th August, p. 15.

LEDA Program CEE (1991) Echanges pour concrétiser sur le tourisme rural. *Buis les Baronnies*, 18–19 janvier, Revue LEDA, London.

LEDIS (1989a) *Agrotourism Initiatives in Greece.* Overseas E58, August

— (1989) *Rural Development in Wales.* Overview B87, August.

Lickorish, L.J. (1989) European tourism 1992: The internal market. *Tourism Management*, June.

Lobry, J. (1986) *Le tourisme vert, un tourisme de liberté.* Paris: TER.

Montemagno, G. (1988) Le développement du tourisme et des loisirs dans les sociétés rurales: possibilités et limites: le cas de l'Italie. Colloquium on 'Tourism and Leisure in the Rural Environment', Sankt Peter-Ording, Schleswig-Holstein, 19th May 1988.

OECD (1980) *The Impact of Tourism on the Environment.* Paris.

— (1988) *Tourism Policy and International Tourism in OECD Countries.* Paris.

Ortuno, M. (1987) *Por un Turismo Limpio.* Madrid.

Planning Practice and Research (1988) Tourism and economic development policy: Policy implications from a case study of Cornwall. *Planning Practice and Research* No. 5, Autumn, 5–18.

— (1988) The growth of leisure: Panacea or political football? *Planning Practice and Research* No 5, Autumn, 14–18.

Planque, V. (1975) La promotion du tourisme vert. Conseil Supérieur du Tourisme, Paris.

Quintrie la Mothe, Th. (1989) Le phénomène de modialisation de l'industrie du tourisme. *Problèmes économiques*, No 2108, 18th January.

Ragu, D. (1988) Les parcs de loisirs: nouveaux pôles de développement local? *Futuribles*, September.

SEATER (1976) Hébergement et restauration en milieu rural. Ministère de l'Industrie, Paris.

Tighe, A.J. (1986) The arts/tourism partnership. *Journal of Travel Research*, Winter.

The tourism industry and the tourism policies of the twelve member states of the community. Department of Economics, Trinity College, Dublin, November 1986.

Tourism is a growth industry. Report No 14 on Tourism, Ministry of Industry, Oslo 1989.

United States Congress (1989) Report of the Federal Task Force on Rural Tourism to the Tourism Policy Council, Washington DC, 25th September 1889.

What are the expectations of the tourists? University of Bergen, 1989.

Tourism Promotion and Potential in Peripheral Areas: The Austrian Case

Richard Hummelbrunner and Ernst Miglbauer
OEAR-Regionalberatung GmbH, Amalienstrasse 68, A-1130 Wien, Austria

The tourist demand for rural areas has accelerated in recent years. However, the supply side remains very much marked by the predominant types of accommodation, while on the other side there is the necessity to provide an increasing range of services to keep visitors interested or attract new clients in rural tourism. From the experience in Austria the key elements of 'new rural tourism' in harmony with man, culture and nature and based upon future market trends are: accommodation which is original and typical for the area; a restaurant sector going for local atmosphere and both seasonal and local specialities; new profiles focusing on specific themes (e.g. health, culture) or groups (e.g. families, the active elderly); a natural and cultural environment which is refined and presented in an appealing and didactic manner, taking the interests of the resident population into account and establishing a new partnership with agriculture. The guiding line for OEAR (Austria's Association for Regional Self-Reliant Development) counselling tourism development in rural areas is built on the locally available resources and potential. OEAR orients itself on three basic principles: environmental soundness, social acceptance and local value-added.

Tourism Development in Rural Areas

Demand for Rural Tourism

Tourism has been one of the most important and fastest growing sectors in Western European economies during the last decades. There can scarcely be any doubt that the demand for tourism will rise steadily in the years to come (Smeral, 1990), and this is all the more likely as it is due to a number of independent factors: improving living standards; new trends in the organisation of working time on an annual basis, allowing for more diversified holiday patterns; longer life expectation and higher incomes for the elderly. The main point is an increasing time-budget for tourism resulting from a reduction in working hours and a strong tendency to use this free time for tourism and leisure activities.

In Germany and Austria the fact that labour as a determining factor in life is declining goes hand in hand with the diffusion of leisure activities in all areas of life (Opaschowski, 1992).

In accordance with the general trend, the development of tourist activities in rural areas has accelerated in recent years. However, there is nothing to say that the predicted increase in leisure will benefit rural tourism in particular. Surveys of holiday preferences among people living in the EEC show that rural tourism is in a mid-way position between seaside holidays on the top and mountains or towns on the other end of the preferences scale. In addition, a 'normal' country-side does not seem to constitute much of a lever for increasing demand, as the majority of tourists are looking for 'natural curiosities', the type of exceptional scenery which only very few rural areas can offer.

Thus rural tourism can and undoubtedly will benefit from the general trend towards more holidays, but this does not necessarily mean that there will be any increase in its relative share (Keller, 1991). However, this somewhat sceptical outlook can be modified by two other factors. The position of rural tourism is much better when only considering secondary holidays. And as splitting appears to be the general trend in the organisation of holidays, meaning several short rather than just a single long holiday, it would appear that rural areas have much to benefit from this trend.

Secondly, a significant and relevant *substitution of clientele* is taking place: the traditional, economically less favoured clientele is declining; rural tourism is gradually losing its image of 'a cheap holiday'. And a new and better-off clientele is becoming increasingly interested in rural tourism (Keller, 1991). It is highly probable that this will become the determining factor for the evolution of tourism in rural areas and will ensure a sound base for substantial growth.

It is both this emergence of a new profile and the *increasing environmental awareness* of holidaymakers, which are becoming the determining factors for the future of rural tourism. The motive 'to get out of polluted areas and into an intact environment' has nearly doubled over the last five years. As an example, nearly 60% of all Germans now claim this as a principal motive. Consequently, the image of a given area is increasingly dominated by the state of its environment, with pollution, traffic and mass tourism as the key negative factors.

Studies carried out in Austria and Germany show that it is the middle and upper socio-professional categories who dominate the sample of those interested in rural tourism, and more than 50 % of these holidaymakers have some higher education. The new potential clientele are the relatively young or middle-aged with a good educational level and family responsibilities. The reasons for choosing rural tourism are: contact with nature, recreation, health, peace and quiet, child-orientation, rural life and contact with local people, and reasonable prices (Hamele, 1991).

Recent Trends on the Supply Side in Rural Areas

The supply side in rural areas remains very much controlled by the *predominant types of accommodation*. Apart from the obvious demand for higher standards of comfort and quality, general dynamic tendencies for the different types of accommodation are:

- The hotel trade is progressing little and has difficulty in adapting to the standards of comfort demanded by the new clientele. Non-classified hotels in rural areas are declining especially.
- Self-catering apartments, cottages and holiday villages are expanding considerably (Smeral, 1990). These types of accommodation are most convenient for families, and they also contribute towards alleviating the work-burden of farm-wives in peak season.
- Boarding with local residents — especially farmers — has been growing rapidly in Europe since the beginning of the 80s, but in German-speaking countries this has long been common practice. In Austria approximately 25% of farms have been receiving tourists for nearly 100 years now. At the

moment, 'farmhouse holidays' contribute to a secure livelihood for 10% of all farms in Austria (Pichler, 1991).

The other important factor is the *necessity to provide an increasing range of services*, to keep visitors interested or to attract new clients in rural tourism (Chassagne, 1991). The idea is to not just provide board and lodging, but to provide a more comprehensive tourism product by offering an attractive range of activities and services.

These developments have important implications for organising the supply-side:

- Activities in rural tourism involve fewer fixed stays, but instead imply constant moves between different forms of accommodation, partners or locations. The supply therefore needs to be *organised in networks* (Greffe, 1991), and the services thus provided are becoming at least as important as accommodation supply.
- The supply needs to be of *increasingly high standard*, e.g. high quality services in terms of information, logistic support or the hire and maintenance of equipment. This requires an adequate level of training in certain fields and adequate pricing of such services.
- In order to ensure quality, suppliers in many cases organise themselves under *defined labels* and make sure they are respected. The organisational initiative may either come from national bodies, regional and local development agencies or associations of the enterprises concerned. The more these organisations are locally rooted, the better it is to ensure and enforce quality standards, due to social control. However, it is this very same force that often makes it difficult to reach a preliminary agreement on these matters.

Setting up new rural tourism systems thus implies organisational difficulties, but it also enables a different economic strategy to be pursued. Economies of scale are at the core of traditional tourism, but the concentrations realised in order to achieve these economies can lead to very grave consequences: escalating costs, environmental damage, the disaffection of tourists, hence problems of amortisation and the setting of exorbitant prices.

Rural tourism, which increasingly consists of services and products jointly provided within a specific location, opens up the possibility to move away from economies of scale and towards *economies of scope* (Greffe, 1991): create additional revenues not by maximising the number of tourists, but rather through maximising the opportunities for them to spend money.

Cooperation becomes a strong condition not only for starting projects in a regional setting, but also for creating a sound environment for investment. Projects have to be big enough to interest banks, but in a small-scale rural setting have to be attractive for the present tourist market. Because banks are hesitant about the small-scale structure of rural tourism, they don't want to invest small amounts of capital in many different projects, it is a challenge for firms and communities to work together in a larger-scale setting, so they may be more interesting and attractive to banks and investors (Speelman, 1991).

The Emergence of a 'New Rural Tourism'

Both the perspectives on the demand side and the trends on the supply side indicate the emergence of a new type of tourism for rural areas, which differs so considerably from the formulas applied so far, that one can truly speak of a 'new rural tourism'. Whereas rural areas in the past were mainly seen as the residual 'hinterland' of tourist centres, 'rurality' is now becoming a new resource and rural areas are in demand for their specific qualities (Keller, 1991).

The *principal considerations* for such a 'new rural tourism' are:

- Rural areas should not try to copy or imitate existing tourist centres; this model is unsuited and financially inaccessible for them. Instead they should innovate in terms of products and their marketing in order to achieve a unique profile, based on their own resources and strengths.
- Rural areas are subject to increasing competition in the tourist market, and neither protectionist measures nor massive subventions can be deployed to prevent this from happening. This situation requires persistent marketing efforts for specific areas or products and the elaboration of a unique selling position.
- Instead of viewing tourism development as an isolated sector, one should opt for a comprehensive and integrated approach to develop a given area or local economy. The focus is to be on the increase of the value-added through the development of corresponding service sectors and economic linkages.
- Use, improve and combine in new, imaginative ways the existing elements instead of focusing on costly investments, e.g. in sports- or leisure-infrastructure. Aim for low investments in board and lodging and make better use of scenic or cultural attractions.
- Give equal attention to the 'software' elements (training, organisation, marketing) instead of being strictly 'hardware' oriented. At least the same amounts should be invested in software areas, namely the human resources and quality-improvement of services.

From the experience in Austria, the following ten elements can be seen as determining *factors for success* of this 'new rural tourism' (Bratl & Miglbauer, 1991):

(1) A sound environment as well as an attractive scenery and cultural heritage are the basic preconditions.
(2) Sufficient entrepreneurial resources, either from private businessmen, enterprising communities, 'social entrepreneurs' or local leaders.
(3) External inputs, including external investments, the creation of necessary infrastructure or the transfer of information and know-how.
(4) A competent and highly motivated steering-group at local level composed of the principal actors and decision makers.
(5) Professional consultancy services to support local initiative with economic, organisational and marketing know-how.
(6) An integrated and participative approach looking for socially acceptable and environmentally sound solutions.

(7) The professional management of projects, enterprises and tourism organisations.
(8) Support by the public sector, especially financial backing for the necessary hardware and software elements.
(9) Partnership and a spirit of cooperation between enterprises, communities and initiatives.
(10) Persistence, patience and time, as tourism development also implies changes in an area's social structure.

The Implementation of Tourism Development in Rural Areas

Strategies

Which are the strategies most likely to succeed? First of all it must be emphasised that there is no 'right' strategy per se, but that the appropriateness of a strategy must always be judged on the area's background. Thus different contexts and situations require different strategies.

At least three *different sets of rural contexts* can be noted (Hummelbrunner, 1991):

- Regions with a tradition in tourism which therefore have the necessary know-how, experiences, organisational structures etc.
- Regions with little tradition in tourism, but considerable potential for development (scenery, cultural heritage, climate etc.)
- Regions with no significant tourist activities and (normally) only average or negligible potential.

Regarding strategies, rural areas have basically *three different options*:

(1) *An endogenous strategy*, focusing primarily on developing and mobilising local resources (natural, physical, financial, human) and potential.
(2) *An exogenous strategy*, focusing on the attraction of external investment to be used as leverage for promoting tourism activities and for overcoming local deficiencies (e.g. capital, know-how). If this strategy is chosen, one has to deal essentially with financiers and/or investors, whose rules and terms for investment in projects need to be taken into account. Trust (in the project and the management) and profit are the most important criteria when considering investments in projects.
As a consequence, it is advisable not to start out by looking for outside capital, but by 'looking inward'. First one has to take a good look at one's area, identify strong and weak points, threats and opportunities as well as local leadership. Only well elaborated projects which demonstrate initiative, self-confidence and the ability to work together will be attractive for outside finance.
(3) *A combined strategy*, which aims at linking endogenous development efforts with the attraction of inward investment, thus hoping to combine the advantages of both of the above mentioned strategies. It is essential to find somebody (ideally a regional development agency) capable of identifying those sectors and areas where local entrepreneurs are not willing or able to offer

desirable products and services, and to orient external investment towards those directions.

External investment is seen as playing a key role to raise the standard of tourism projects to national and international level. The public sector should also be considered as an important investment partner, because tourism development in many cases demands enormous investments in infrastructure. In peripheral rural areas, these costs cannot be met by the local population, who normally lack funding and are also not able to draw sufficiently high benefits from these investments. Nor are private investors willing to take over these costs, so they must be financed or at least subsidised by the public sector.

How do these strategy options which have been outlined relate to the different types of context? The following matrix shows how the appropriateness and thus the chances for success can be estimated. It is primarily based on Austrian experience, but will equally hold true for large parts of rural Europe.

	Strategies		
Areas	*Endogenous*	*Exogenous*	*Combined*
Tourism tradition	+ +	+ -	+ +
Less tradition	+	+	+ +
No tradition	- -	+ +	-

Key: + appropriate; + + very appropriate; + - average; - less suited; - - not suited

Thus endogenous strategies have the highest chances for success in areas with a tradition in tourism, which already have the necessary human resources at their disposal, namely enough people with sufficient qualifications and experience. In areas with less tradition in tourism both endogenous and exogenous strategies are difficult to operate and a combined strategy seems most appropriate. And in areas without a tradition in tourism only an exogenous strategy has a chance to succeed — provided that the existing potential is attractive enough for external investors. On the other hand, this strategy is less suited for areas with a tradition in tourism, because external investment can collide with the interests of existing entrepreneurs and can even impede the mobilisation of local resources.

The chances of success of these different strategies are essentially *a question of time*. Based on the experience in Austria, endogenous strategies can be realised in areas with a tourism tradition within 3–5 years, whereas it takes nearly double this time (7–10 years) in areas with less tradition. When external investors in a combined strategy act as 'accelerators' of the development process, the respective time spans range from 2–3 years (tradition in tourism) to 5–7 years (less tradition).

Structures

It is not a question of selecting *one* single structure, but of the proper combination of structures and organisations at different levels and with different tasks (Hummelbrunner, 1991):

- *Local and regional tourism organisations*: their task is to develop and market the tourism products of a given area. They should be able to bring the respective actors together in a cooperative spirit and to assist entrepreneurs and initiatives.
- *Support structures at the local level*: they should facilitate the transfer of information (e.g. on market needs and possibilities, potential investors, financing), but also give advice and support for businesses, projects and development programmes.
- *'Top down' support structures*: they should assist with the proper marketing of tourism products in external areas as well as support the activities on local and regional levels (by financial means, services, personnel etc.).

In most rural areas, especially those on the periphery, it is the establishment of local tourism organisations which seems most urgent: they are capable of properly representing and articulating local interest in tourism development. It is through them that individual interests and projects can be brought to form a coherent strategy and can be communicated to external investors, public administration etc. It is also the prerequisite for the above mentioned 'new rural tourism', which is the result of a highly demanding and complex development process. It therefore calls for a highly participative and integrated development approach.

A second element of priority is the setting up of local consultancy structures. Easily accessible and affordable consulting services have become a key element for the promotion of local economic development in rural Austria — and in many other parts of Europe. The possibilities on the travel market have been outlined above, but it is not easy for rural peripheries to tap this potential. Consultancy services with a long-term and activating approach are an important element to bridge this gap.

Tourism Development in Austrian Rural Peripheries

Key elements of the approach of OEAR

The OEAR (Austrian Association for Regional Self-Reliant Development) is a non-governmental intermediary agency for the promotion and support of endogenous local development. It operates primarily in declining rural areas and the focus is on consultancy services in key sectors for economic renewal, namely agriculture, tourism, small-scale industry and renewable energy resources. Those services are provided through a network of 60 'regional advisors' working out of ten regional offices in peripheral (mainly rural) regions throughout Austria.

The guiding line for OEAR's work is the endogenous development strategy, building on — and preserving — the locally available resources and potential, be it human, physical or financial. In counselling tourism development, OEAR orients itself on three *basic principles* (OEAR-Regionalberatung GmbH, 1991):

Environmental soundness:

- Conservation of existing natural environments
- Limitation of 'accessibility'-infrastructure
- Gradual and controlled improvement of tourist infrastructures

- Use of local architectural styles

Social acceptance:

- Preservation of socio-cultural identity
- Promotion of locally rooted culture (crafts, artistic expression and customs)
- Up-grading of job-quality in tourism
- Democratic control of tourism development
- Prevalence of the resident population's interests (over those of the tourist population)

Local value-added:

- Best use of existing capacities and local resources
- Wide-spread distribution of economic benefits through the creation of local small and medium enterprises (SMEs) and their cooperation
- Cooperation between tourism, agriculture and SMEs
- Integrated development instead of economic mono-structures

In order to be able to effectively promote tourism in peripheral areas, the profile of the tourism products to be established should build on their *key strengths*:

- Attractive natural environments with considerable ecological qualities, that offer a potential for emotional/aesthetic perception and experiences.
- An original regional culture, which is marked both by regional history and — often — by interesting cross-cultural relationships.
- Systems of land-use and agriculture, which not only have had a long-term (and visible) impact on landscapes and culture, but are still in use, preserving nature and its resources.

If it is to be successful, tourism should not only be in accordance with an area's strengths, it must also build on *future market and cultural trends* (Hamele, 1991):

- a growing environmental awareness;
- an individualisation and a sense for originality and openness for new experiences;
- a growing desire for highly original, spontaneous and non-'prefabricated' experiences and encounters;
- a growing desire to be able to participate and to be integrated.

This development of the tourist's motives is not always unambiguous. Different and sometimes contradictory holiday expectations are to be found in the post-modern holidaymaker — pleasure and unassumingness (modesty) at the same time (Smeral, 1990).

And last, but not least, one should take into account — and possibly reduce — the key weaknesses of peripheral areas:

- lack of local entrepreneurs and entrepreneurship;
- lack of know-how and experience;
- lack of capital;
- very limited basic infra-structure.

The promotion of local entrepreneurship focuses on a principal deficiency and bottle-neck. But whereas the lack of capital can be overcome through loans and subsidies, it is much harder to substitute a lack in experience in isolated regions. The tools to tackle this problem are training and a maximum involvement of local residents with tourists and tourism activities, even if its only very small scale.

The *key elements of 'new rural tourism'* in harmony with man and nature and based upon future market trends as well as the strengths and weaknesses of peripheral areas are:

- Accommodation which is original and typical for the area, thus presenting an appealing alternative to standard-type accommodation found elsewhere.
- A restaurant sector going for local atmosphere, seasonal and local specialities, especially highlighting fresh and high-quality food produced in the area.
- New profiles focusing on specific themes (e.g. health, culture) or groups (e.g. families, young people, active elderly), creating new and interesting mixtures of tradition and innovation.
- A natural and cultural environment which is refined and presented to the visitors in an appealing and didactic manner, to please their curiosity and desire for emotional experiences.
- Taking the interests of the resident population into account and establishing a new partnership with agriculture by promoting its role as preserver of a sound natural environment and as supplier of high-quality food products.

Main challenges/problems facing rural tourism areas

A very short survey of problem areas:

Conflict between short-term visitors and holiday makers:

This phenomenon is most acute in rural areas that have major attractions for short-terms visits (e.g. natural parks, castles, lakes, skiing centres). The main problems are traffic circulation and assuring zones of remoteness/quiet. Partial solutions are being tried out occasionally (Hopfenbeck & Zimmer, 1993) but lack diffusion and exchange of experiences.

Development planning in rural tourism:

How to handle the — often conflicting — interests of different types of tourists (e.g. short-term visitors, holiday makers, users of apartments and secondary homes) by means of planning. What methods to apply, so as to reconcile the demands from these different target groups?

Conflicts arising from widespread use of new 'trend-sports':

Notably in summer-time, new trend-sports are becoming a more and more familiar feature, even in remote rural areas, which offer the necessary terrain and ample natural resources for the practice of sports such as: mountain biking, rafting, hydro-speed, para-gliding, bungee-jumping, free climbing. They bring with them a series ecological, social and ethical conflicts (Miglbauer, 1992) and increasing tension between 'conserving' and 'utilising' natural resources.

Regulatory measures in rural tourism:

Such measures have the objective to limit the impact on environment, residents and other guests. At present they are mainly used in national parks, but in the future they will be widely applied in rural areas. Examples are: regulations at local/regional (e.g. no-trespassing, seasonal limitations) or national level (e.g. zoning), 'ticket into nature'-pricing, information and didactic measures (Hopfenbeck & Zimmer, 1993).

Cost-benefit relations in rural tourism:

Strategies that were considered 'patent-solutions' (e.g. holiday villages) are turning out to be increasingly problematic. Increasing criticism is being voiced because of generally degrading cost–benefit relations in rural areas (Gaudriault, 1991). How do we to achieve tourism development which is both sustainable and (cost)-effective?

References

Bratl, H. and Miglbauer, E. (1991) Oekologisch und kulturell profilierter 'Neuer Landtourismus'. OECD-Symposium 'Rural Tourism in a Changing Europe'. Wien: OEAR-Regionalberatung.

Chassagne, M.E. (1991) Tourisme et développmement local: sortir des illusions. *Actes des recontres des Baronnies,* janvier 1991.

Gaudriault, M.C. (1991) Le tourisme rural en France: réveiller la Belle au bois dormant. *Actes des rencontres des Baronnies,* janvier 1991.

Greffe, X. (1991) Trois idées fixes en guise de synthèse de la conference. *Actes des rencontres Baronnies,* janvier 1991.

Hamele, H. (1991) Die Nachfrage nach laendlichem Tourismus. OECD-Symposium 'Rural Tourism in a Changing Europe'. Wien: OEAR-Regionalberatung.

Hopfenbeck, W. and Zimmer, P. (1993) *Umweltorientiertes Tourismusmanagement.* Landsberg/Lech: Verlag moderne industrie.

Hummelbrunner, R. (1991) Abschiessende Bemerkungen. OECD-Symposium 'Rural Tourism in a Changing Europe'. Wien: OEAR-Regionalberatung.

Keller, P. (1991) Tourismus und Regionalentwicklung. OECD-Symposium 'Rural Tourism in a Changing Europe'. Wien: OEAR Regionalberatung.

Miglbauer, E. (1992) Neue sommer-trendsportarten — konflikte und loesungsansätze. In W. Pillman and S. Predl (eds) *Strategies for Reducing the Environmental Impact of Tourism.* Wien.

OEAR-Regionalberatung GmbH (1991) *Tourismus — Zukunfts-Gemeinden — Neue Angebote — Tourismus im Einklang.* Wien: OEAR Regionalberatung.

Opaschowski, H. (1992) *Freizeit 2001.* Hamburg: BAT -Freizeitforschungsinstitut.

Pichler, G. (1991) Neue Ansaetze im Baeuerlichen Tourismus. OECD-Symposium 'Rural Tourism in a Changing Europe'. Wien: OEAR-Regionalberatung.

Smeral, E. (1990) *Tourismus 2000.* Wien: Signum Verlag.

Speelman, K. (1991) Löst Auslandskapital alle Probleme? OECD-Symposium 'Rural Tourism in a Changing Europe'. Wien: OEAR-Regionalberatung.

Rural Tourism as a Factor in Rural Community Economic Development for Economies in Transition

Agnes Gannon
20 Glenbrook Park, Rathfarnham, Dublin 14, Ireland

This paper examines how community resources can be mobilised to help rural communities make the transition from agrarian to diversified and sustainable economies. It discusses the concept of the Global Management Approach used by the Food and Agriculture Organization of the United Nations in Europe, and shows how this approach can be used to develop sustainable rural tourism, paying close attention to implementation issues.

Introduction

Structural change in economies is causing severe stress in many rural areas especially for those areas with a heavy dependence on agriculture. This is particularly the case for countries which are changing from a 'command' to a market economy. However, Central and Eastern Europe do not have the monopoly of 'economies in transition'. The whole of Europe is in the process of changing its economic and social character and its strategic orientation. All countries in Europe have problem areas or regions where farm revenues are decreasing, farm land values changing and unemployment rising, leading to an exodus of productive forces and demographic imbalances in rural areas. Consequently there is a need to change the crisis combat ability of these economies in order to help them adapt to changing technologies and market place conditions as well as helping them diversify their economic base to meet the changes.

The real challenge to policy makers, leaders and rural communities is to find innovative responses to current problems and thus facilitate the transition and adaptation of economies to new economic and social structures.

Many rural areas with economies in transition contain valuable reservoirs of skills and abilities as well as space and other physical resources which, if mobilised, offer potential for endogenous and self-reliant development. For many countries this adjustment and development process requires a renewal policy, partnership between Government sectors, between public and private sectors and above all between the people of the community who are the main actors in the development drama.

This paper proposes to look at how community resources can be diversified and developed in order to facilitate the transition and adaptation to new economic and social structures which will contribute to long-term sustainability. The paper focuses on two main aspects: (1) Rural diversification for sustainable economic community development and (2) Rural tourism as a factor in the economic development of rural communities.

Rural Diversification for Sustainable Economic Community Development

The economic crisis of rural areas can be said to be the consequence of a complex set of forces which have resulted in the decline in both job opportunities and in wealth creation for rural communities. This has been mainly due to the fact that:

- Food requirements which are dictated by economic efficiency are being produced on smaller areas of land by fewer people.
- There is a shift from labour intensive to knowledge intensive systems in the manufacturing industry.
- The service sector development has been mostly urban located and hence many rural opportunities have been unexploited.

In promoting long-term sustainable economies for rural areas agriculture is still deemed an important engine of economic growth. There can be no rural life without agriculture — it is the cornerstone of rural economic development. Therefore, the maintenance of a healthy agricultural economy remains a critical component of any comprehensive effort aimed at rural economic development.

Efficiency, effectiveness and competitiveness are imperative to agricultural development of the future. There is a continuing need for the development and adoption of improved production-practices, management and marketing technology and in particular for entrepreneurial practices which provide scope for flexibility and creativity in the diversification of the economic base.

In the past, many rural employment opportunities were unexploited because agriculture tended to decouple itself from the manufacturing industries by its preoccupation with the traditional production of more and more food. Consequently the possibilities of creating endogenous enterprises relating to the pharmaceutical, fibre and service systems of agriculture were neglected. Therefore, future development in agriculture must *not* only look at economies of scale but also economies of scope in agriculture. It is necessary to recognise the virtues of diversification for agricultural development and for economic development as distinct from economic growth. Keane & Quinn (1990) argue that there is a clear difference — economic growth implies more of the same while economic development is a process of structural change implying something different.

According to Lohmöller (1990) the shaping of future rural economies through diversification depends on new ways of doing things, i.e. seeing things differently in order to do things differently. The scope is tremendous and future success in rural development will depend largely on economies of scope. This requires a whole new approach to a thinking process which is entrepreneurial related.

In the past the limits of the entrepreneurial concept in the agriculture sector was 'production' which stemmed from the traditional thinking of food security. This level of thinking collapsed with market saturation. Then the concept was extended to price which became the central focus of agriculture policies. This concept contributed to a positive development in agricultural income *but* its focus had to shift from rising prices to how to avoid falling prices. This led to developing the concept of optimising the cost structure within the enterprise. Since not every farmer could manage this approach the rationalisation resulted in an increase in part-time farmers.

However, all of these concepts were not enough to ensure survival of an enterprise or business within the constantly changing macro environment. Related problems can be complex and interlocked and consequently can be caught up in a whole network of interdependencies. Solving such problems requires an understanding of systems management and the realisation that there is no effect without a side effect. Therefore an entrepreneurial concept encompassing systems management was deemed necessary to manage the complexity of the systems involved. This is known as the Global Management Approach (GMA).[1]

The GMA is an integrated management approach which helps to channel community energy into a progressive self-help process thereby enabling people to analyse their own situations, generate solutions to their problems, plan their own developments and evaluate results. It provides the tools for examining each situation in a systemic and systematic way. This is done by using a holistic systems approach based on clearly defined strategy and planning methodology which details the sequence of events leading to implementation and the evaluation of results. GMA responds to the needs of simultaneous planning of multipurpose activities required in integrated rural management strategies. Because of its holistic approach GMA ensures that agricultural changes are paralleled by the development of the secondary and tertiary sector activities and the appropriate social and administrative institutions. For idea generation and problem solving the GMA uses the 'Concept Mapping' process. This is a method of visually representing concepts relating to a given issue. It provides for a democratic decision making structure within the dynamics of the group process which removes the possibilities of threat and domination by members of the group and it also provides an atmosphere for mutual understanding and confidence. The process ensures increased group participation where all ideas are shared and no idea is lost. It ensures shared leadership, innovation and flexibility as well as group ownership for problems, solutions, projects and products thus enabling the development of informed local control.

The GMA encompasses an entrepreneurial approach to development which is now being recognised as a key factor of modern economic development. Essential elements for entrepreneurial development are:

- Flexibility — which is necessary to enable the adjustment to new and unforeseen situations.
- Creativity — which provides for the allocation of resources from less productive use to more productive use.

The ability to change and be innovative is central to the entrepreneurial process in rural economic development. This can be particularly effective if organised through group entrepreneurship.

Petrin (1991) argues that group entrepreneurship is key to the economic development of rural areas. It provides social assets, a range of skills, the possibility for a balanced management team, combined strengths as well as psychological support which is particularly necessary at 'start-up' time. Development of entrepreneurship can be stimulated by supportive institutions, by deliberate action to stimulate change and also by providing support to capable individuals and groups.

An important element in the development of entrepreneurship is training at the different education levels. Petrin (1990) states that special training is important because in most countries the education system does not emphasise entrepreneurial ideas or abilities. Some vocational institutions touch on this to a very limited extent but in general the school system clearly educates to salaried positions. It can therefore be argued that education systems should set their future focus on educating 'to make a job rather than educating to take a job'.

Within the rural development agenda small-scale industries/enterprises have acquired a new relevance based on the assumption that many of them offer a competitive advantage in the development of rural regions. Local entrepreneurs of small enterprises can with modest outlay contribute considerably to economic growth because they supply smaller markets, demand relatively small amounts of capital, use local resources and raw materials and do not require costly and sophisticated infrastructure. For this reason many countries can stimulate rural development by stimulating the development of a strategy for small-scale enterprises. Rural tourism falls within this category. By its very nature it is the core of economic community development as it can build on the existing economic base of rural resources for the development of tourism related businesses.

Rural Tourism as a Factor in Economic Development of Rural Communities

The rapid growth and the accompanying visibility of tourism has raised the awareness of many policy makers and local leaders of the importance of rural tourism as a community development vehicle. However, in many rural areas the rural tourism concept has done little for community development. On the contrary it has remained unchanged and is still operating as 'Holidays' with the market dominated by an image of accommodation, supported by vague statements of 'fresh air, healthy living and lots of things to do in the country'. In the meantime tourism in general has grown into a major economic sector which has become extremely competitive and is dominated by communication and promotion techniques. Compared to this general development rural tourism has, for the most part, remained an 'amateur production'.

The blurred image model of rural tourism has now reached its limits. If it is to reap its full potential as a vector of a new phase of economic development for rural communities it will have to adapt to current market mechanisms to comply with customer needs. No longer should the product be built around 'selling what we can produce' *but* 'producing what we can sell' with a customer orientation should be the motto.

The current dynamics of tourism development requires the development of services, activities and the environment as well as quality accommodation. Therefore, if economies in transition or if communities in the process of development are to effectively diversify the rural economic base through tourism and if they are to compete effectively in tomorrow's tourism industry they will need to understand and consider the essential issues for rural tourism development. According to the experience of the author the following seven issues are deemed important in the development of rural tourism projects:

(1) Economic ambition.

(2) Authenticity.
(3) Benefits and costs of tourism.
(4) Tourism as part of a strategic development plan.
(5) Developing and delivering quality products.
(6) Training.
(7) Future trends.

Economic ambition

This involves a sense of development by rural communities, that is, that rural communities are prepared to take their future into their own hands and promote a long-term policy of economic diversification which can be suitably adapted through small-scale industrialisation for the creation of new employment opportunities. Rural tourism by virtue of its 'knock-on' effect offers a major potential for this development.

By definition rural tourism is a collection of businesses that create sales of goods and services to tourists. These sales include spending by tourists at local businesses which normally service the local residents as well as spending at businesses that are primarily tourist targeted. Businesses which service the tourist are important to the development of the total tourism package and to the development of the area. Too often the development of business and trade are neglected while product providers concentrate their efforts on accommodation and food. Failure to give equal priority to the development or expansion of relevant businesses results in a less attractive tourism package and also misses the opportunity to create jobs.

Economic ambition behoves rural community members as it empowers them to create, develop and to control the rural tourism product and other forms of diversification, the direct or indirect benefits of which, can be shared with the community members and can also benefit the area and the region.

Authenticity

Authenticity provides the competitive edge for the tourism product. Its main ingredients are: genuine quality, originality, uniqueness, a sense of place and a pride of place. It is manifested in the lifestyle and the life quality of the community and in particular it relates to the behaviour, integrity, friendship and sincerity of the people who live in and develop the community. It is also influenced by the physical, social and economic nature of the rural area, i.e. space, heritage, agriculture, landscape, services and the significant historical and cultural events of the area.

The components of the tourism product such as the people, the accommodation, the activities, the attractions and the available services should reflect authenticity in the quality of life of the area. Authenticity rates very highly with tourists on the 'Benefits' scale, and is a major incentive in bringing tourists back to the area.

Benefits and costs of tourism

Rural tourism is a socio-economic environmental business and can therefore generate both advantages and disadvantages in the economic, social and environmental spheres. To avoid growth potential problems, a careful assessment of both advantages and disadvantages should be made at the preplanning stage by the community and the actors concerned. The following outline of Benefits and Costs (adapted after Richardson, 1991) can assist in the decision making process:

Economic benefits

Rural tourism:

- Brings more money particularly foreign currency to the economy.
- Creates jobs and increases family and community income.
- Helps to diversify and stabilise the rural economy.
- Provides a broader business base for the community and creates an opportunity for attracting other businesses and small industries.
- Provides the opportunity for innovation and creativity.
- Provides the support for existing businesses and services.
- Helps to develop local craft and trade.

Economic costs

Rural tourism:

- Incurs developmental and marketing costs.
- Involves risk for the operator and community.
- Places demands on public services.
- May only give part-time employment.
- Is an image industry and is therefore very sensitive to the macro environmental conditions and forces outside direct community control.
- May increase cost of living for community residents because of inflation of property and goods and services costs.

Social benefits

Rural tourism:

- Fosters a pride of place and a sense of place.
- Helps to build up a community infrastructure.
- Provides the opportunities for cultural exchange.
- Promotes a team community spirit particularly through the development of cultural and entertainment activities.
- Creates conditions for safeguarding and enhancing local cultural identities.

Social costs

Rural tourism may cause:

- The introduction of conflicting ideas and styles into the community.
- The increase of crime.
- Overcrowding and congestion.
- Infringement on privacy in households and in the sharing of important community resources with outsiders.
- Institutional, community and family jealousies, as all may not share the benefit equally.

Environmental benefits

Rural tourism:

- Can be a key factor in revitalising the natural, cultural and historical resources of a rural region.
- Promotes and encourages village renewal and a cleaner countryside.
- Can foster the preservation and conservation of worthwhile resources.

Environmental costs

Rural tourism:

- May downgrade the quality of natural and historic areas through increased tourism and tourist behaviour.
- May increase noise and litter pollution.

Rural tourism as part of a strategic economic plan

Rural tourism is not the universal solution to the ailments of the rural world. It is one enterprise which can possibly have significant economic 'spin-off' effects which could help to slow down or in some cases counterbalance the disintegration of the local fabric. But to be a totally effective rural development tool it must be part of a strategic economic development plan, that is a plan which will help to combat agriculture underemployment and rural unemployment generally.

Such a plan should encompass the social, economic, environmental, physical and administrative development of an area and should aim at diversifying the economic base through complimentary activities. These activities could include services, craft and maintenance industries, auxiliary farm and non farm products, small farming properly networked, small manufacturing businesses and wholesale and retail trade. Through a diversification strategy, with a broad level of community participation and with local initiative, greater security is provided for the community than through single sector development. Single economic development is synonymous with insecurity in rural development terms.

Developing and delivering a quality tourism product

Developing and delivering a quality product is what the whole tourism business is about. The two main components are Production and Marketing both of which depend on quality for their life line. In developing and marketing tourism it is necessary to understand the features which distinguish it from other products:

- It is a service industry where the customer comes to the product.
- The product is sold on the basis of many intangibles i.e. emotional benefits such as experiencing the wonders of nature, nostalgia, friendship, relaxation, adventure, fresh air, space, entertainment and recreation.
- Most customers rely on marketing messages as a basis for holiday decision and therefore buy the product without first seeing it.
- Purchases are made on a high level of trust in quality delivery.

Developing the product

The establishment of an entrepreneurial group which is capable of addressing the collective needs of the community is essential in the initial stages of 'start-up'. The group should have a clear vision of its task. It should be trained in human development, leadership, group dynamics, management, problem solving and in how to interact and network with other organisations and agencies. Such training is a basic necessity for the sustained progress and development work of the group and it should help to prevent competitive rivalries that can waste time, effort and money and can destroy the energy, leadership and ownership potential of a group.

Creating a vision for the group should be a lead task. Such a vision should reflect not only the direction of tourism but should also focus on the desired development results of the community. This vision should be clearly understood by all members of the group.

Leadership is key to group cohesiveness and it is also necessary for group management. The concept of shared leadership, which recognises that power is greater when united through a collective unit than when splintered, is recommended for group entrepreneurship. Members of the group need to be able to come to agreement on common goals and in solving common problems. Shared leadership and ownership facilitates this development. Ownership is part of the stakeholder concept, where individual members of the group in addressing problems or issues have commitment to the resolution rather than the transference of responsibilities to others.

Development starts with marketing because the motor of marketing is customer needs (Drucker, 1986). Therefore, product providers must understand the customer and the tourism features and the benefits which they seek. They must build their product from an assessment of the communities strengths and weaknesses, from market research and from the needs of the tourism environment.

Micro-marketing which focuses on providing different brands for different niches is now in vogue. Present trends in tourism indicate that markets must be targeted and that rural community tourism should not try to be all things to all people. Instead it should be identified with unique features which should set it apart from the competition. Marketing objectives must be in line with the target market while product development objectives must be in line with what the community and what the area has to offer, that the customers need.

In setting up a strategy for developing and delivering the tourism product it is important to realise that it must be user oriented and not seller oriented. It should conceptualise the product from the customers' point of view and should reflect market trends.

Critical questions for a development/delivery strategy

Answers to the following key questions should provide a basis for a development and a delivery strategy:

- What does the community and the area have that visitors want?
- What attractions, activities and accommodation facilities are necessary to entice visitors to come and stay?
- What has the community to offer in attractions, activities and accommodation and what is required?
- What are the opportunities for creating additional revenue in the area.
- What services and access facilities are available and what is required?
- What is the image of the area?
- What and where is the competition?
- What are the current markets?
- What are the trends that effect the tourism industry? (Gannon, 1992)

The overriding question is: 'Has the community product package sufficient *drawing power* to bring visitors to the area who will generate a significant economic income and who will return to the area?' Remember that the package must offer Quality, Uniqueness, Authenticity and Drawing Power in order to remain on the cutting edge of competition.

Training

Maintaining sustainable rural tourism requires quality and professionalism which in turn requires training for rural tourism entrepreneurs. Professional training for the task should aim at providing the entrepreneur with an understanding of the following six competencies:

- A global management approach to rural development.
- Rural tourism development.
- Group development and dynamics particularly for entrepreneurial groups.
- Product development.
- Promotion and marketing.
- Customer care and after sales service. (Gannon, 1992)

In addition an understanding of how to package the elements of these six competencies into a strategic management plan is necessary.

Future trends

There is a continuing need to update the rural tourism product and future trends should signpost the direction i.e.:

- Growth rates in tourism will continue to expand.
- Experiencing and experience have a top priority for vacations.
- Vacation is increasingly determined by recreation activities.
- Visitors bring their environmental ethics with them — there will be greater emphasis on environmentally friendly destinations.
- There is increasing interest in authenticity.
- There is a tendency to take health benefits for granted i.e. tourists expect menu options and fitness exercise options etc.
- Social tourism which puts emphasis on human relations, freedom, self-fulfilment and conviviality is gaining in popularity.
- Senior citizens are becoming an increasingly important target group.

Conclusions

Initiating/developing a rural community economic development programme is challenging and rewarding particularly if it encourages local participation and is based on the premise that the local community has the power and the ability to impact on its own future developments and direction.

The potential of a sustainable community programme is limited only by the vision and drive of the people driving the process. If a community can develop a vision of where it wants to be, then steps to get there can be developed and put in place.

Tourism offers the rural world a second chance. Its future and sustainability will depend on a quality product which promotes human values, authenticity and ecological stability. Future investment must therefore keep the landscape intact; foster a healthy social structure of the local communities; promote optimal quality of life and relaxation for visitors and increase the potential for added value in the tourism industry. Rural tourism can no longer afford a second-class image. It has all the ingredients and potential for a first-class sustainable product. This can be made a universality by all concerned.

Notes

1. The GMA system has been successfully used by the Food and Agriculture Organization of the United Nations (FAO) in the development and implementation of rural development and rural tourism projects in many European countries.

References

Bornet, Bernard (1992) Rural tourism as a socioeconomic factor stabilising the rural population: An appropriate alternative to the pressure felt in overcrowded tourist areas. Council of Europe Colloquy, Palermo, 1992.

Drucker, Peter (1986) *Innovation and Entrepreneurship*. Suffolk, UK: The Chauser Press Ltd.

Gannon, A. (1990) A global management approach to rural tourism development. FAO: 3rd International Summer School on Rural Development, Vienna, 1990.

— (1991) Rural revitalisation. DPI Seminar, Queensland, Australia, 1991.

— (1989) Improving the vitality of rural areas through community business initiative, IRD International, Workshop, Tel Aviv, Israel.

— (1991) Group entrepreneurship. FAO: 4th International Summer School on Rural Development through Entrepreneurship and Village Renewal, Heimai Centre, Finland.

— (1992) Training for rural tourism. FAO Workshop: Rural Tourism Development, Rinn Bei Innsbruck, December 1992.

— (1993) Concept mapping in developing group projects. FAO Summer School, Bled, Slovania.

Keane, M.J. and Quinn, J. (1990) Rural development and rural tourism. University College Galway.

Lohmöller, G. (1992) Concept for the development of entrepreneurial activities in rural areas. 6th Session of the ECA Working Party on Women and the Agricultural Family in Rural Development, Innsbruck.

Koth, B., Kreag, G. and Sem, J. (1991) A training guide for rural tourism development. Minnesota Extension Service, University of Minnesota, USA.

Mesl, Mateja (1992) How rural tourism can assist economics in transition. FAO Workshop on Rural Tourism Development, Rinn Bei Innsbruck, Austria.

Petrin, T. (1990) Entrepreneurship and small business development. FAO/ECA 5th Session of the Working Party of Women and the Agricultural Family in Rural Development, Prague.

— (1991) Rural development through entrepreneurship. FAO: 4th International Summer School, Rural Development Through Entrepreneurship — Village Renewal, Heimai Centre, Finland.

Richardson, S.L. (1991) *Colorado Community Tourism Action Guide*. University of Colorado.

Shaffer, R.V. and Palmer, G. (1987) *Economic Development for Rural Revitalisation: A Handbook*.

Marketing for Agri-tourism in Austria: Strategy and Realisation in a Highly Developed Tourist Destination

Hans Embacher
Austrian Farm Holidays Association, Hardtgasse 19, A-1190 Vienna, Austria

Tourism plays a dominant role in the Austrian economy. A total of approximately 109,000 rooms are offered on 21,000 Austrian farms. Thus the farmers represent one sixth of the total Austrian supply of tourist beds. In recent years the holiday farmers have organised their representation and marketing more professionally, first on a regional level, then in the provinces. Finally in 1991 the Austrian Farm Holidays Association was founded to represent the interests of the farmers on a nationwide level. A positive development towards 'farm holidays' as a brand name has begun generating an additional source of income for farmers and enabling tourists to spend interesting and valuable holidays on an Austrian farm. The paper outlines the situation of farm holidays in Austria (statistics, etc.), the tasks of the organisation to promote farm holidays on each level as well as the nationwide strategies of the Federal Association (common logotype, uniform system of quality standards, specialised products). Finally some conclusions are drawn from the work in this intersection of tourism and farming. One such conclusion is that farm holidays are ecological in providing an environmentally and socially compatible form of tourism, supporting small-scale enterprises and integrating the local people.

The Role of Tourism in Austria

Tourism and the leisure industry are vitally important to the Austrian economy and play a key role in economic growth, employment and the balance of trade. With expected foreign exchange earnings of AS 170 billion in 1992, tourism is unquestionably one of the cornerstones of the Austrian economy.

The position of the Austrian tourism industry in the world is exceptional: with a share of world GDP of only 0.7%, with 0.9% of the world's population and a share of visible exports of 1.7%, the small country of Austria ranks sixth in the world in terms of vacation traffic and fifth in terms of tourism foreign exchange earnings. With foreign exchange earnings from tourism of approximately AS 21,000 per head of population (in 1992) Austria has by far the most intensive level of tourism amongst the Western industrialised nations (second is Switzerland with 12,000 AS foreign exchange earnings per head of population). The most intensive tourist region in Austria is the Tyrol, with 71.3 visitor-nights (12.6 arrivals) per head of population.

8.5% of the GDP and 20% of all export earnings (goods and services) derive from tourism. Some 400,000 persons earn their living from tourism. Tourism in Austria also has a considerable effect on the regional balance. Regions with little or no industry (mountains, etc.) are in general more attractive areas for tourism. It is estimated that the total turnover of tourism and tourism related activities will have reached 400 billion AS by 1992. This is equivalent to 15% of the Austrian GDP.

In Austria 83,300 tourist enterprises offer some 1.2 million tourist beds. This results in an average size of 14.4 beds per enterprise. Thus, tourism in Austria is characterised by a small scale structure. In 1992 the tourist industry reached an all time high of 25.8 million arrivals and 130.5 million visitor-nights. The average length of stay is 5.05 days/person, and the average load factor is 106 days per bed/year. The total number of beds has been about the same for the last 15 years, but the structure of the accommodation has changed: beds in commercial enterprises with a higher quality standard (4/5 stars) and apartments have increased, while beds in private accommodation (maximum 10 beds) have decreased considerably.

Table 1 Country of origin of tourists in Austria (based on visitor-nights in 1992)

Germany	49.6 %
Austria	23.5 %
Netherlands	6.9 %
Great Britain	3.1 %
Italy	2.8 %
Belgium/Lux.	2.3%
Others	11.8 %

Table 1 shows that 73.1% of all visitor-nights comprise guests from either Germany or Austria. This reflects the dependency of Austrian tourism on the German as well as on the home market. Indeed the — for long neglected — home market has proved to be of growing importance over recent years.

Agri-tourism in Austria

Since the first beginnings of Austrian tourism, private homes and farmhouses had let rooms to travellers. Up to the 1960s and 70s this was known as 'Sommerfrische' (summer recreation) — a popular form of spending the summer holidays in the countryside.

Contributions of farm holidays to tourism and farming in Austria

Farm holidays are an important element to sustain rural life in general and the farming sector in particular.

The farming sector

(i) Farm holidays are a chance for a second or third economic leg for farms to stand on besides, for example, cattle and timber. In this way, fluctuations in prices in the meat or timber market can be balanced out. The farmers remain self employed and may be in a position to remain working solely on the farm, for example without having to commute to a factory in the next town. This is a key element of the farmers' identity.

(ii) Farm holidays are also a chance for the direct sale of farming produce. First, farming products are used for the breakfast as well as for meals, and this can contribute to a positive image of farming produce in general. Secondly, farmers who sell their products directly to consumers often use their guests as an interesting distribution channel and sometimes even inform their guests (those who do not live too far away) when freshly harvested goods are available (biological farms in particular use direct mailings).

(iii) At present in some sectors of the farming industry there are considerable problems and limitations (overproduction, GATT restrictions, etc.). Tourism, however, is considered to be a growth sector and the demand for sustainable tourism in particular seems to have good growth potential in the long run. Thus, income from tourism seems to be a reliable source of income independent from development in the agricultural sector.

Tourism

(i) Guests on a farm have the chance of getting to know nature, people and the region usually in a small-scale enterprise with personal service by the landlord and landlady. Thus a good deal of 'everyday culture' of the country and region may be experienced.

(ii) Farm holidays can contribute considerably to the economic survival of some farms (especially in the mountain regions with much tourism). The small-scale structure is a key feature of Austrian agriculture. Especially in the Alps, only farms up to a certain size can be run effectively by a family, and bigger farms can only be worked with considerable use of machines, which can hardly be used in steeper regions. Tourists, however, like and expect a landscape made by the farmers. If too many farmers quit, this could considerably affect the landscape which research suggests itself attracts many tourists.

(iii) Farmers who rent out rooms link tourism together with agriculture. A good deal of the tourists' activities take place on farmers' land (skiing, cross country skiing, walking, mountain biking, etc.). This is a source of potential conflict between the two sectors, even more so when tourism seems to do well and the farmers have problems. Thus good communication between the two sectors is vital to the healthy development of tourism in the long run. In this respect 'holiday-farmers' have an important role to play.

(iv) Farm holidays are a contribution to sustainable tourism. Farm holidays are a gradually grown regional offer which needs very little specially built infrastructure. In most cases existing resources in the farmhouses are done up and rented out to the tourists. Thus, existing potential in terms of labour and buildings is used. Economically farmers do not depend solely on tourism, and therefore they can withstand weaker seasons or a recession in tourism more easily than someone who totally depends on tourism. Farm holidays also help ease regional congestion during the main season as many farms are in more remote areas or in regions with little tourism. In some Austrian regions tourism reaches limits as far as acceptance by the local population is concerned. The more locals benefit from tourism usually the better the acceptance of tourists. Farm holidays also need hardly any labour from outside. Thus farm holidays are a form of tourism provided by the locals and where the benefits mainly remain in the region.

Table 2 Farm holidays in Austria: enterprises, rooms and beds on farms

Region	*All categories*				*'Private on farm'* *	
	1980		*1990*		*1990*	
	Farms	*Rooms***	*Farms*	*Rooms***	*Farms*	*Rooms*
Burgenland	980	4,765	672	3,504	170	1,356
Carinthia	4,704	29,109	3,468	19,522	1,735	13,747
Lower Austria	2,675	11,470	1,733	7,691	634	4,479
Upper Austria	3,285	15,958	2,346	10,725	1,073	8,183
Salzburg	3,796	19,717	3,059	15,929	2,291	17,286
Styria	4,421	21,517	3,022	14,183	1,664	11,953
Tyrol	7,511	41,118	6,006	33,061	3,991	29,951
Vorarlberg	1,568	6,840	970	4,281	401	2,688
Austria	**28,958**	**150,618**	**21,276**	**108,896**	**11,959**	**89,643**

Source: Austrian Central Statistics Office

* The category 'Private on farm' includes tourist enterprises with a maximum of 10 beds (apartments are not included). The enterprises in this category are included in regular annual statistics. There is no regular statistical data available on farms with more than 10 beds and on farms which offer apartments, apart from the agricultural census every 10 years.

** The agricultural census counts only the number of enterprises as well as the number of rooms in farmhouses, not the number of beds. However, it is estimated that a total of 220,000 beds is offered on farms.

Analysis of supply

In total, some 21,000 farmers offer around 109,000 rooms (including bedrooms in apartments), which means that 7.5% of all Austrian farmers offer tourist accommodation. This represents roughly 18% of the total Austrian supply of tourist beds. Of this total, about two thirds are farmers in mountainous regions.

The number of enterprises (26%) as well as the number of beds offered (28%) has decreased considerably from 1980 to 1990 (Austrian Central Statistics Office, agricultural census, carried out every 10 years).

Detailed statistics are only available on farms with up to 10 beds and which do not offer apartments (according to the Austrian laws these enterprises are included in the category 'private on farm'). This category has 11,000 enterprises with 84,000 beds. Thus the available statistics reflect the — on average — 'weakest part' of farm holidays as apartments in general show the highest growth rates and bigger enterprises are generally more professional in their marketing approach. Nevertheless, the available statistics may function as an indicator for the development of farm tourism in Austria.

Figures 1 and 2 show that the number of enterprises and beds in this category is decreasing steadily. Some farmers find it more attractive to offer apartments, some grow and offer more than 10 beds and are therefore no longer in the above category. Some others however end their tourist activity or fall out of the market.

There are several reasons for this decline in farms involved in tourism:

- The overall number of farms in Austria is decreasing.

- Involvement in tourism without a certain investment in time and quality of accommodation is not really worthwhile economically.
- Farming and tourism in combination is a lot of work for the farmers, especially for those who also work outside as employees.
- A certain investment has to be made to improve the standard of the accommodation and to stay competitive. For example, there is very little demand for rooms without a shower and WC.

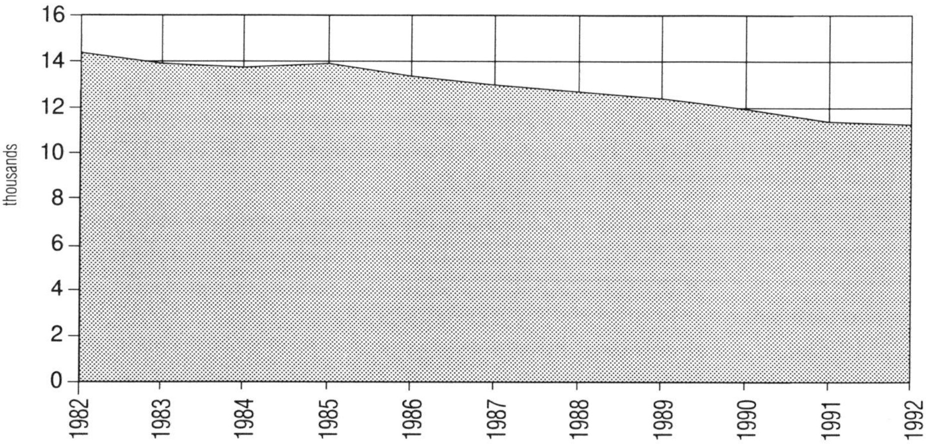

Figure 1 Number of farm holiday enterprises, 1982–1992. (This is based on the 'Private on farm' category which excludes enterprises with more than 10 beds and any apartments.)

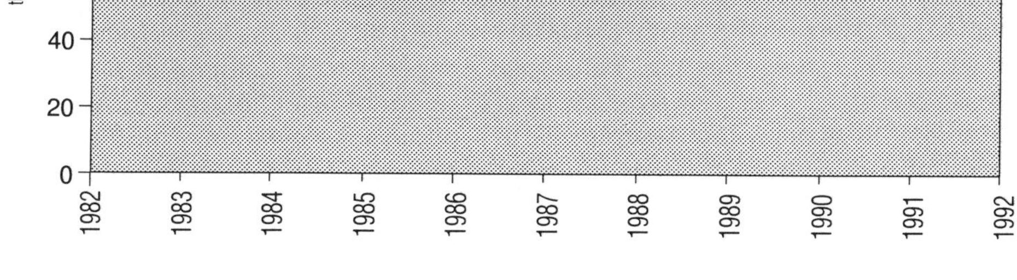

Figure 2 Number of farm holiday beds, 1982–1992. (This is based on the 'Private on farm' category which excludes enterprises with more than 10 beds and any apartments.)

- In many cases the young generation which takes over the farm do not permanently want people in the house. Often they have experienced how there is very little privacy for the family.
- During the 1960s and 70s tourism was booming in Austria and there was strong demand for private accommodation. During the 80s the supply of hotel-beds has improved and tourists could afford higher standards of accommodation. Consequently, the importance of private accommodation, including farm holidays, was declining. From 1980 to 1992 the number of private beds has declined by 33% and is still declining steadily. However, in this sector there is strong demand for self catering apartments.

The following information reflects the situation of the 3,300 members of the Austrian Farm Holidays Association (out of a total of around 21,000 farmers involved in tourism). It may be assumed that these members on average put more emphasis on tourism than other farmers who rent rooms to tourists (at least they use more marketing channels than the others):

- 58% of the total capacity are rooms with shower/bath and WC, or apartments. It becomes increasingly difficult to sell rooms with a lower standard. However, 40% of the total capacity still do not include these facilities.
- The farmers mainly offer bed and breakfast.
- Half board arrangements have to include regional cuisine, as much as possible with products from the farm. For some farms which are too far away from the next village, offering half board is a necessity.
- Apartments on farms are becoming increasingly popular. Usually no service is provided, so it is more easily handled by the farmers. However, there is growing demand for apartments with breakfast or half board.
- 'Tourist products' are being developed to meet specific demand and for marketing to specified target groups e.g. packages for fishing (farms have to be directly by the lake!), riding, cross country skiing, or health holidays on a farm.
- An increasingly important aspect to be integrated in farm holidays is the direct selling of farm produce to the visitors — from dairy produce and foodstuff to local handicrafts.

Analysis of demand

As already pointed out, there is statistical data available only on farmhouses with a capacity of up to 10 beds. Thus the following statistics represent a segment of some 11,000 farms with a total of 84,000 beds. All other farms are included and counted in other categories (e.g. 'Pensionen' or apartments) — therefore we have no exact statistical overview over the development of farm holidays in Austria as a whole. In general the Austrian Farm Holidays Association would define farms with a maximum capacity of up to 30 beds as farms for 'farm holidays', but they would exclude a farmer who runs a 60-bed hotel — which is not unusual in certain parts of the country.

From 1981 to 1985 the bednights on farms have declined from about 6 million to 5 million bednights (-17%). Since 1985 the number of bednights has stagnated at around 5 million (the number of beds has declined by some 20% from 1980 to 1990). However, in 1991 there was an increase of more than 10% to 5.4 million

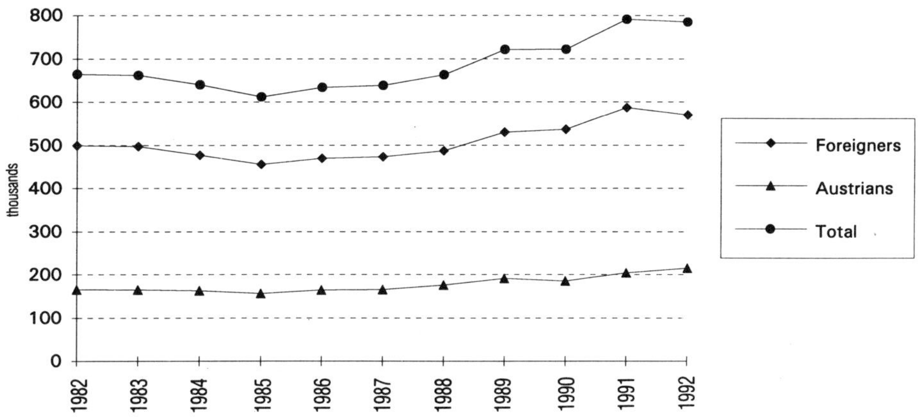

Figure 3 Tourist arrivals on farms, 1982–1992. (This is based on the 'private on farm' category which excludes enterprises with more than 10 beds and any apartments.)

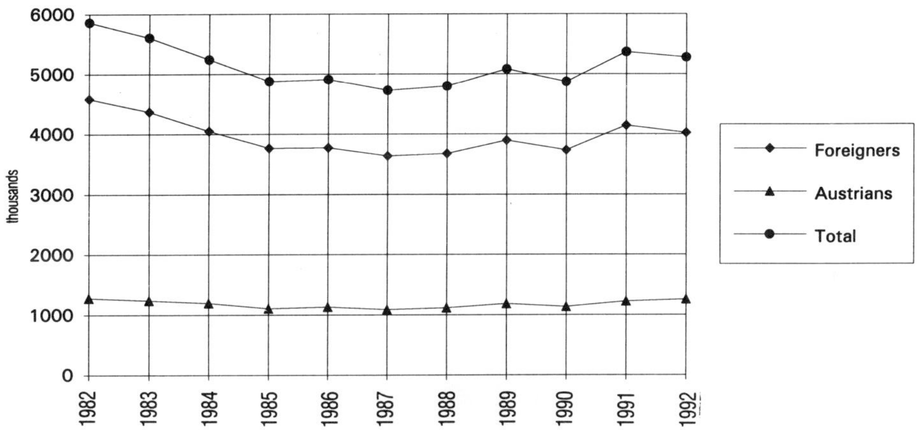

Figure 4 Overnight stays on farms, 1982–92. (This is based on the 'private on farm' category which excludes enterprises with more than 10 beds and any apart-

bednights. However, as indicated earlier, this figure includes only farms with a maximum of 10 beds. There is reason to believe that farms with higher capacity have developed more favourably as they find it easier to adapt to changing trends due to reasons of finance, more interest and greater emphasis on tourism, etc.). It is also known from surveys that the members of the Austrian Farm Holidays Association on average do perform better than the 'statistical average'.

Seasonality

44% of all annual bednights are during the months of July and August (Figure 5). This too indicates the high proportion of families who tend to travel during their children's school holidays.

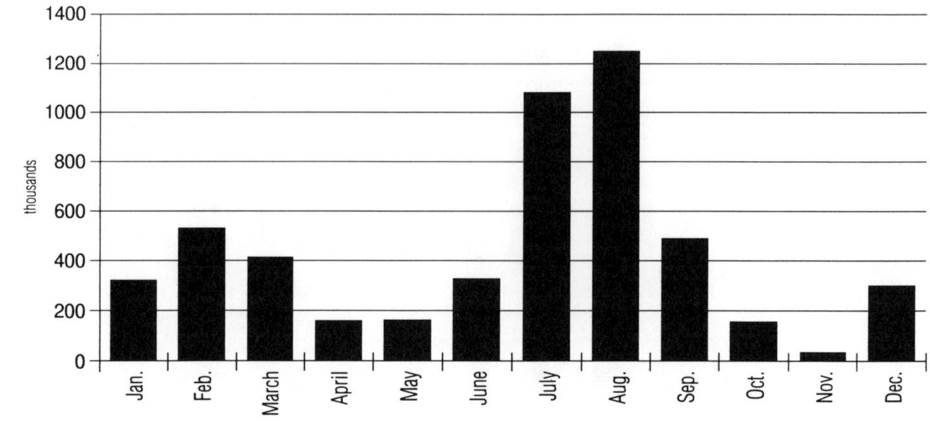

Figure 5 Average overnight stays by month, 1991–1992. (This is based on the 'private on farm' category which excludes enterprises with more than 10 beds and any apartments.)

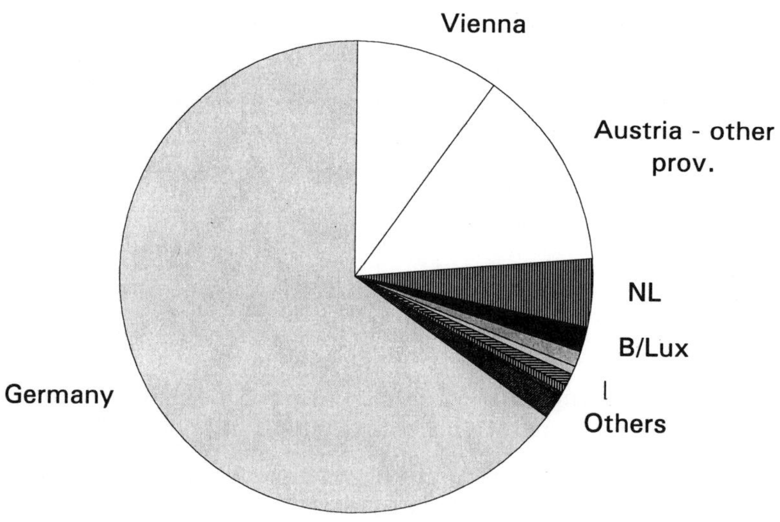

Figure 6 Place of origin of visitors staying overnight, 1992. (This is based on the 'private on farm' category which excludes enterprises with more than 10 beds and any apartments.)

Countries of origin

64% of visitors staying overnight on the farms in 1992 were from Germany, and 23% were from Austria (Figure 6). Hence almost 9 out of 10 guests on farms are either Germans or Austrians. The reason for this high proportion may be that

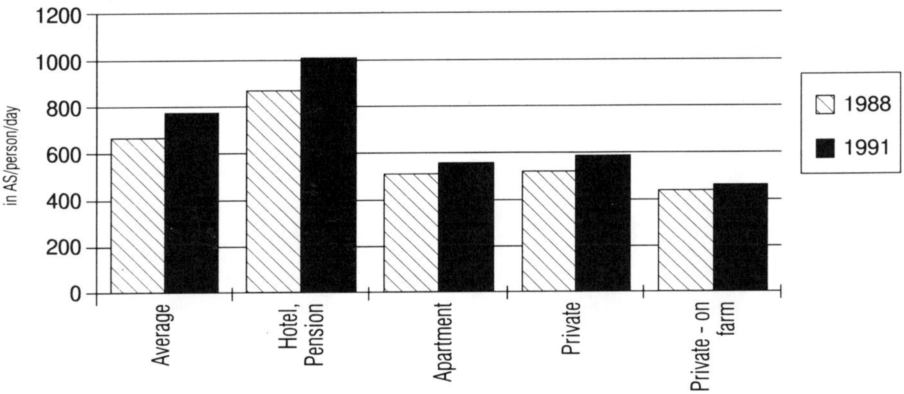

Figure 7 Daily expenditure in the summer seasons, 1988 and 1991)

guests on farms are mainly families with children who do not like to travel long distances, who like the cultural familiarity, and who make direct reservations. The highest proportion of guests come in contact with the farm through word-of-mouth recommendation by friends.

In the future the main emphasis will remain on the German speaking markets (Austria, Germany, Switzerland, Netherlands and Belgium) as this is also in the interest of the farmers. However, certain provinces will also present their product outside the German speaking areas (e.g. the Tyrol has achieved considerable growth rates in France, Carinthia and Upper Austria in Italy).

The results in Figure 7 indicate that on average holidaymakers on farms spend less money than people who spend their vacation in other forms of accommodation. The expenditure of tourists in other forms of accommodation has grown relatively faster from 1988 to 1991 (Private on farm: +20 AS; Private: +70 AS; Hotels: + 140 AS; Average: +100 AS). This indicates that farm holidays are clearly positioned as a low price product.

The main target group of the future will remain families with small children (farms are ideally suited to these families). 38.7% of all German families with small children say that holidays on a farm would be an ideal form of vacation for them, but only 2% of this potential effectively spend their holidays on a farm. Other interested groups may be ecologically sensitive tourists (especially those interested in biological farms as well as farming produce), special interest holidaymakers (riding, fishing, health, etc.) as well as maybe to a certain extent elderly people (See Figure 8).

The Austrian Farm Holidays Association

Until the 1970s there was no organisation to coordinate the work of farmers who offer accommodation. For a long time farm holidays were considered as a cheap low quality product with little potential in the market. The first organisation to promote farm holidays was a regional pool of farmers in the Salzkammergut — a tourist region situated in the provinces of Salzburg and Upper Austria — founded in 1971. The first provincial Farm Holidays Association was founded

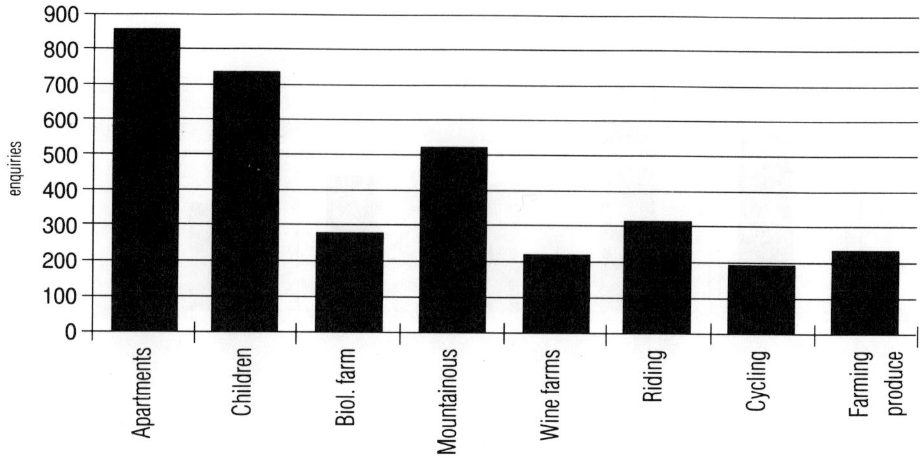

Figure 8 Special interests identified in a survey of enquiries, 1992. (Source: 1,500 enquiries to the Federal Association, 1992.)

in Styria in 1972, the most recent one was in the province of Salzburg in 1992. The Austrian Farm Holidays Association ('Bundesverband für Urlaub am Bauernhof in Österreich') was created by the eight provincial Associations as a coordinating and integrating body in May 1991.

Organisation

Out of a total of about 21,000 farmers who rent rooms to tourists, some 3,300 farmers are voluntary members of the Austrian Farm Holidays Association. They represent a total of about 45,000 beds. These farmers pay a contribution (on the basis of 10 beds) of, on average, about AS 1000 per annum — farms with higher capacity pay higher fees. Between 1988 and 1990 a concept was worked out to establish an organisational structure to promote farm holidays throughout the country. This structure includes three organisational levels (destination/region, province, nationwide) and this structure was established in 1992 (see Figure 9).

Tasks of each organisational level

(1) Main tasks of farmers as landlords

- Product design.
- Provision of a good quality tourist product.
- Provision of precise information for guests.
- Co-operation and involvement at all levels.
- Pre-defined quality guarantee and adequate farm management.

(2) Main tasks of regional pools or guest circles

- Organising exchange of experiences among farmers.
- Organising training, consultancy, and assistance to individual farmers.
- Organising booking system.
- Organising advertising on regional basis.
- Assisting provincial associations in the respective region.

Farmers as Landlords for Farm Holidays
3,300 farmers (out of a total of some 21,000) are members of a regional pool responsible to one of the eight provincial associations.
Local and Regional Pools or 'Guest Circles'
In some provinces on a district administrative level (e.g. in the Tyrol) there are eight regional pools. Usually the work of these pools is coordinated by one person.
Eight Provincial Associations
These are found in all provinces except Vienna. Their offices are mostly within the premises of the Chamber of Agriculture, from which the associations have evolved. Usually one person or a manager and secretary coordinate their work.
Federal Organisation (The Austrian Farm Holidays Association)
Established in May 1991, this is the coordinating body of the eight provincial associations. The office is at present within the premises of the Austrian National Tourist Organisation. Currently there is a manager and a secretary.

Figure 9 Organisational structure for the Austrian Farm Holidays Association

(3) Main tasks of provincial associations

- Organising and optimising marketing (in co-operation with all levels).
- Organising and executing booking system at the provincial level.
- Setting up and promotion of regional guest circles.
- Maintaining contacts with provincial tourism organisations, chambers of agriculture and provincial governments, and with the federal level.
- Exerting an influence on preparation of offers (promotion, consultancy, vocational training, quality).

(4) Main tasks of federal organisation

- 'Preparing the ground' for agritourism in Austria on a national level.
- Exploiting all possibilities and opportunities for promoting farm holidays at the federal level.
- Representation of the interests of farm holidays on a national ('lobbying') and international basis.
- Maintaining contacts with the Austrian National Tourist Office, the Federal Ministry of Agriculture and Forestry, the Federal Ministry for Economic Affairs, the Chambers of Agriculture, training institutes, etc.
- Assisting, motivating and coordinating the activities of the eight provincial associations.
- Steering national and international marketing.
- Central information office for farm holidays in Austria (no booking).
- Basic market research.

Finances

The existing Farm Holidays Associations are sponsored by:

- Member fees of the farmers.

- Subsidies and support by the Chamber of Agriculture (in most cases the offices of the Associations are provided by the Chambers).
- Support by the Ministry of Agriculture and Forestry with special projects (e.g. subsidies for brochures and leaflets, sponsoring of certain projects , training of staff, etc.).
- Support by the Regional/Provincial/National Tourism Organisations, especially with the production and distribution of advertising material, presence at trade fairs in Austria and abroad, and with know-how.

In 1993 the federal association worked with a budget of around 3 million AS.

Marketing

The overall marketing strategies are discussed later. The organisation tries to support and promote direct contacts between (potential) guests and farmers as individuality and personal contacts are major strengths of the product. This is also expressed in the catalogues (at the provincial/regional level — there is no nationwide catalogue with all farms), in which every farm is presented with a photograph and short description, address and telephone number. Farm holidays are presented at trade fairs in Austria and abroad, in advertising campaigns (in newspapers and magazines, but with very limited budgets) as well as in PR activities. Besides the special catalogue 'Farm Holidays in ...' the farms are usually also presented in the leaflets and brochures of the resorts. A further main feature of all marketing strategies to promote farm holidays are close links with the tourist organisations on all levels.

Booking/reservation

Some associations (e.g. Vorarlberg, Tyrol, Upper Austria, and one region in Lower Austria) cooperate with travel agencies which enables them to make direct reservations on a provincial level. The Tyrolean association is a member of 'Autriche pro France' and 'Austria per l'Italia' to promote farm holidays in these countries. The Tyroleans have also integrated the direct selling of farming products in their farm holiday catalogue to create a whole (some) product.

Overall marketing strategies

- More activities at all levels of marketing.
- Making use of all possible synergy effects.
- Better co-operation with existing tourism organisations — without losing individuality!
- Creation of an independent product philosophy — an 'active market policy'.
- (Quality) classification of the product.
- Emphasising regional particularities.
- Raising price levels in general and extending price margins (higher quality products at higher prices).
- A family-orientated pricing policy.
- More market differentiation of offers to target groups.
- Promotion of direct contacts between guests and landlords.
- Preparation of cooperation with travel agencies and service centres (on a limited scale).
- Development of a network between the regions/provinces.

- Emphasising PR activities.
- Creation of corporate identity/corporate design.
- Systematic market research.

Implementation of strategy

The Association is working nationwide in the following areas:

(i) *Strengthening the self-esteem of the farmers*

The aim is to develop more professional products and higher prices with consequently increased profitability. Activities include:

- Supplements on 'farm holidays' in nationwide agro-magazines.
- PR in agro, tourism and public media.
- Participation in seminars, workshops, discussions; and represention of 'farm holiday' interests there.

(ii) *Creating a basis for joint marketing*

It is assumed that demand for farm holidays will increase considerably in the future. The product has to be prepared to make sure that this will bring mainly positive effects to the farmers. Activities include:

- Establishing a common logotype for farm holidays in Austria to be used at all levels. This logo should stand for good quality.
- Implementing a uniform system of quality standards (from two to four 'flowers' similar to the system of 'stars' used by hotels). At present some provinces have no system, some use three, others four classes with quite different criteria, which makes it difficult for the consumer to identify the product which suits him best. The new catalogue of criteria has been worked out by farmers, consultants and members of the organisation in a preparatory and discussion process of one year and is now available as a brochure. Besides furniture and equipment the criteria include aspects like 'farm quality', 'experience quality', information for guests as well as ecological points. The farmers are also obliged to offer their guests at least three home made products. The nationwide classification starts in 1993 in four provinces and is to be completed within two years. It is organised by the provincial associations and is valid for a maximum of four years (according to the catalogue-cycle in every province).
- Support for farmers and associations with the specialisation of the product(s). For this purpose a marketing handbook has been initiated together with a bank. The handbook is accompanied by a series of seminars on marketing for farm holidays in every province (also organised and sponsored by the bank).

(iii) *Image work*

The aim is to create an image of an attractive, interesting tourist product which is of good quality and adequately (not cheaply) priced. Activities include:

- Image brochure.
- Adverts in family magazines.
- Regular PR.
- Television programmes (e.g. a children's programme from the farm).

- Presentations at workshops and trade fairs (mainly in German speaking regions).

(iv) *Improving information and motivation*

This should be of farmers and staff on all levels. Activities include:

- Seminars for staff and farmers.
- Magazine for the members (planned for the autumn, 1993).
- Regular talks and discussions with farmers.

Training

Staff training

Staff training for the organisation through seminars and courses is organised and subsidised by:

- The Ministry of Agriculture and Forestry.
- The Chambers of Agriculture (Provinces).
- The federal Farm Holidays Association.

The themes are general tourism, trends, marketing and sales, investments, and public relations,etc.

Training of the farmers

This is usually arranged on a local or regional level where there is a better knowledge of interests and demand and the closer contact ensures higher attendance.

- The farmers usually have to pay a small contribution.
- These seminars are often sponsored by local banks or tourism organisations.

The themes include the pricing policy, guest correspondence (writing letters, telephone training) and certain skills like weaving or embroidery in order to pass this knowledge on in special courses for guests.

Conclusions

Tourism has a long history in Austria. Since the beginnings, farmhouses and private homes have been a popular part of the Austrian tourist product. In the 1970s and 80s the tourist needs and consequently demand has changed considerably. Most visitors to Austria do not come to the country any more simply because they look for recreation. Today they look for more 'intelligent' products; they clearly want to see their personal benefit, such as why they should go to this resort, to this specific hotel or why they should stay at a farmhouse.

During the 1980s the total bednights on Austrian farms declined, but since 1990 there has been a considerable increase in bednights. In recent years farmers who rent rooms have organised their representation and marketing more professionally — first on a regional level, then in the provinces. Finally, the Austrian Farm Holidays Association has been founded to represent the interests of the farmers on a nationwide level.

Today many concepts and ideas concerning the organisation as well as the product are being considered or are in the planning stage. But a positive development towards 'farm holidays' as a brand name has begun to provide the farmers with an interesting source of income and to allow guests to have interesting and valuable holidays on an Austrian farm.

With the opening of borders to Eastern Europe, Austria's position within Europe may be newly defined politically (Austria intends to join the EEC) and in tourism terms, on the crossroads between North and South, East and West. For the farmers this may well be an additional chance as they offer relatively cheap accommodation, compared to hotels. It is hoped this chance will be used by some of them who offer lower standards, to improve the quality of their accommodation in order to stand a good chance in the bigger, more open tourism markets of the future — which will bring new opportunities as well as new threats.

For the farmers, the future will require a clear decision as to whether they want to continue being a by-product within the tourism policy of their country, their region or their resort, or if they are prepared to play an active, creative role in the development of tourism on every level! The setting up of the presented organisational structure throughout Austria is a further step towards more professional work.

Some general conclusions may be drawn about farm holidays in Austria:

- There is a good level of market potential for farm holidays.
- It is in the interest of everybody involved in tourism to promote farm holidays.
- Farm holidays are ecological — it is an environmentally and socially compatible form of tourism, it supports small scale enterprises and integrates the local people.
- The small scale enterprises (with a capacity of around 10 beds) are in particular need of support by a strong professional organisation.
- New product ideas can only be successfully developed in close harmony with the farmers as they often tend to regard new developments with scepticism.
- As a basis for successful marketing it is vital to improve the self confidence of many farmers who rent out rooms — to convince them that they offer an interesting product which is of high value to their guests, especially to families.
- 'Tourism and Farming' will be a theme of the future and the Farm Holidays Association could contribute to establish good relations between these two important branches of the economy.

Some final points should also be highlighted about the necessity for cooperative marketing to make agri-tourism sustainable.

The holiday farms in Austria are a high number of small scale enterprises. This results in a lack of management and marketing know-how coupled with limited financial and marketing potential in order to address more sensitive groups of consumers/holidaymakers. Therefore the argument for farm holidays has for long been that they were the cheapest form of accommodation on the market. This image should gradually be changed towards the valuable elements of holidays in the environment of a farm (e.g. to experience farming life, animals, work on a farm, closeness to nature in the seasonal cycle, healthy food, etc.) which are especially exciting for children who grow up in urban areas. In the future the pedagogical effect of time spent on a 'living' farm will become more important. Besides the chance of experiencing an interesting vacation, the tourists can be made more sensitive to the challenges of the future for farming

Although the small scale of the businesses may be a weakness in marketing terms, it is one of the major strengths in terms of the special 'holiday experience' on the farm. Thus the objective of an organisation is to balance out the marketing limitations and to give the farmers support wherever possible and sensible. This support seems necessary to make sure that the farmers will effectively benefit from the existing market trends and conditions, which within a competitive tourist market like Austria favour an attractive, sustainable, small scale form of tourism.

Sustainable Tourism Systems: The Example of Sustainable Rural Tourism in Sweden

Lars Aronsson
University of Karlstad, Department of Social Sciences, Section for Geography and Tourism, Box 9501, 650 09 Karlstad, Sweden

The aim of the paper is to provide an overview of what is required of a sustainable tourism system with its two main aspects, demand and supply, and the meeting-places between them. The term tourism refers to leisure travel away from home. As regards the demand side, the paper gives an overview of the motives for and restrictions on travel; on the supply side it considers resources for tourism and tourism products. The question of what is sustainable tourism development is examined, with mass tourism being contrasted with small scale tourism and, in particular, ecological tourism. Further, tourists' energy consumption on long distance journeys and tourism's local resource consumption are also discussed. The concept of meeting-place, classified in a number of categories, is used to link demand and supply within the tourism system. The term refers to the encounter between the tourist and the local environment. Sustainable development, which is a normative concept, has several dimensions. Conditions of the following type are generally emphasised in definitions of sustainable tourism development: Sustainable tourism development should entail a long-perspective for the use of resources in tourism production; A form of tourism should be developed that contributes to creating equality and economic and social welfare for the local community; Tourism development should make careful use of the natural and cultural conditions, including the built environment; Positive tourism development in line with the above points should not be a burden on other people and areas or on coming generations.

Introduction

Background

Tourism is today generally considered a major industry in the world. The Swedish Tourist Board (1991a) estimated that the turnover from tourism in Sweden in 1990 was approximately one hundred milliard SEK and that approximately 235,000 people were engaged in the industry full-time or part-time throughout the country.

In many regions in Sweden tourism grew in importance during the 1980s. The early 1990s were marked by stagnation and recession, partly due to the general recession. However, from the winter season 1992–93 tourism in Sweden has once again started to increase, partly as a result of the devaluation of the Swedish crown, which has made it cheaper for foreign tourists to visit Sweden and more expensive for Swedish tourists to travel abroad.

Both as regards the number of visitors and from an economic viewpoint, tourism is most significant in the major cities and in certain coastal (including islands) and mountain areas (see Figure 1).

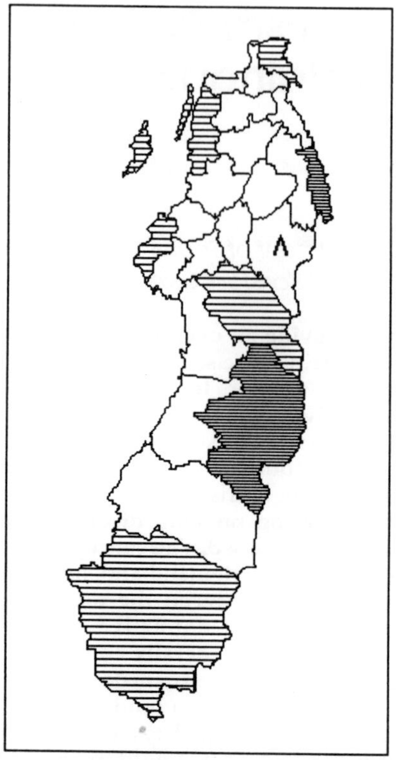

Figure 1 The map shows the relative importance of tourism in the Swedish Counties in 1989. Counties with a turnover from tourism between 11,000 SEK to 20,000 SEK per inhabitant are shaded lighter and counties with a turnover of more than 20,000 SEK per inhabitant are shaded darker. The county of Värmland, referred to later in the text, is marked with V. (*Source*: The map is made on the basis of information from the Swedish Tourist Board, 1991b.)

The development of tourism in an area has a number of consequences for that area. In the following discussion the emphasis will be primarily on the normative concept of sustainable tourism development. Sustainable tourism development has several essential dimensions. Initially, I would like to mention two.

The first concerns the fact that all exploitation for tourism and all travel has an impact on the environment. It is thus a question of establishing acceptable limits for the load/strain which we would wish to allow. One example of environmental impact is the pollution that results from the consumption of energy on long journeys. Are there different types of tourist journeys and are some more acceptable from the energy viewpoint than others? The second dimension is that sustainable tourism should provide a good source of income for and support the culture of the residents in an area. This means that I include economic, social and cultural aspects in the concept and this, in its turn, places demands on, for instance, the form and extent of tourism and on the rate of development.

Aim

The aim of the paper is to provide an overview of what is required of a sustainable tourism system with its two main aspects, demand and supply, and the meeting-places between them. It is also my intention to problematise issues concerning sustainable tourism. The empirical observations are based on material from my previous research (Aronsson, 1989, 1993) and on studies of the literature. In the paper I attempt to achieve a general approach to sustainable tourism development. At the same time, it is difficult to avoid the stamp of the rich world's perspective on development. This bias is also reinforced by the fact that my own empirical material on which the discussion is based is primarily drawn from rural areas in Sweden.

The Content of the Tourism System

What is tourism?

There many different definitions of tourism. Each author defines tourism to suit his purpose in studying the phenomenon (see, for instance, Ogilvie, 1933; Cohen, 1974).

Mathieson & Wall (1982) show how complex the concept of tourism is and indicate the problems involved:

> Tourism is the temporary movement of people to destinations outside their normal places of work and residence, the activities undertaken during their stay in those destinations, and the facilities created to cater to their needs. The study of tourism is the study of people away from their usual habitat, of the establishments which respond to the requirements of travellers, and of the impacts that they have on the economic, physical and social well-being of their hosts. It involves the motivations and experiences of the tourists, the expectations of and adjustments made by residents of reception areas, and the roles played by the numerous agencies and institutions which intercede between them. (p. 1)

I adopt Mathieson & Wall's standpoint as regards the delimitation of tourism. To put it briefly, I use the term tourism to refer to leisure travel to destinations away from the home area. Leisure refers to the time that remains after work and household duties etc. have been performed. The concept is thus limited in time and the spatial dimension is not valid until a person chooses to perform certain activities in his leisure time in the local or non-local environment. However, there are many dimensions to leisure, some of which are difficult to measure scientifically (Swedish Tourist Board, 1986).

The tourism system has several components: demand in the tourism-generating area, supply in the host area and the journey there and back. I shall discuss aspects of all three in the following sections. The tourism system is constrained by economic, political, social, cultural and physical conditions which affect the extent, content and structure of the system in various ways.

The demand for tourist travel

There are different theories as to why people make tourist trips. The various assumptions may be classified in two main groups. The first is an escape hypothesis, that by going on a tourist trip we escape from a drab and sometimes alienated everyday environment. This hypothesis does not necessarily entail a negative attitude to the tourist but can be interpreted as implying that we all need a break from the daily round to 'recharge our batteries' (Krippendorf, 1989). The other, which I term the educational hypothesis, is clearly positively charged and assumes that we seek experiences, knowledge etc. (Karlsson, 1994).

A basic condition for the attempt to achieve sustainable forms of tourism lies in the awareness and attitudes of the individual. What is required is knowledge of environmental issues and an environment-friendly approach.

When studying motives for tourist travel in more detail, we must be aware that typologies always reduce reality. There is a whole set of typologies of tourists, including ones that have been developed for the market segmentation of tourists or designed to elucidate their behaviour. Few of them, however, say much about the impact of sustainable tourism or the demand for resources during the tourist trip (Svalastog, 1992). Lowyck *et al.* (in Johnsson & Thomas, 1992) have critically analysed several different classifications of tourists.

However, since social differences, in the form, for instance, of life mode and life cycle factors, determine where one goes, how one travels and how long one stays in a place and also affect the impact of tourism on the area visited, it is important to differentiate between tourists on this basis. There is a dialectic relationship and mutual influence between the everyday, i.e. conditions at home, and the non-everyday, i.e. the tourist trip.

Before travelling, the tourist must fulfil a number of conditions or have certain resources. Apart from motive, he must have time at his disposal and money for his board and lodging. Finally, travel requires some form of transport, which determines the range in time-space. Thus right from the start there are a number of restrictions on movement.

Finally we must remember that the major part of the total demand and implementation of tourist travel is by people in the rich world, either through travel in our own economic and cultural sphere, or through travel to poor countries.

The supply for tourism

Resources for tourism

Resource is a concept that is used in many different contexts and there is no unambiguous definition. The following discussion concerns area-based resources rather than the tourist's individual resources.

Zimmermann (1972) gives a broad definition of a resource as something that at a certain point in time fulfils a function for people. Each function generally has a purpose, e.g. to satisfy a need. A resource may be seen as a means of achieving certain goals and it thereby reflects people's values. According to Zimmermann's definition there must be a demand before a resource can be utilised for production. At the same time there are many restrictions which prevent a product being consumed by everybody. It should, therefore, be made clear that resources for the production of tourism are to be seen in a dynamic perspective where, among other things, values in the demand change over time.

As Svalastog (1985, 1993) points out, there are, with regard to location, two types of resources for the production of tourism, resource-based and market-based. Typical of the resource-based resources is that they depend on natural and cultural resources, the latter relying on material and immaterial conditions. Market-based location is based on the principle of proximity to the user and includes business and professional travel. Both resource-based and market-based location for tourism production presuppose further resources in the form of, for instance, knowledge and technical development, which are manifested in the infrastructure of society and the structure of the tourist industry, among other things. Apart from these two basic forms, Svalastog (1985) mentions network-based travel, which consists of visits to relatives and friends. Resource-based and network-based travel consist of leisure travel where the latter is non-commercial.

Clawson & Knecht (1966) have made a classification of area-based resources for outdoor recreation which fits in with the above reasoning. This classification sees resources as a continuum extending from intensive utilisation/exploitation to untouched nature. The two poles of the continuum are:

- *User-oriented resources*. The resources are based on available resources close to densely populated areas. They are exploited intensively and exposed to great user pressure. They are often artificial products demanding capital-intensive investment.
- *Resource-based production*. The resource is seen as outstanding for the demands of tourism. Its quality is thus of the greatest importance. Developing it requires relative little capital investment and it is generally based on the natural or cultural product itself. At the same time, the resource is crucial for the profile of the whole area. It is often situated at a distance from densely populated areas.

The emphasis in the discussion of sustainable tourism development is on resource-based travel.

Tourism products

An area's resources for tourism can be utilised to build up its supply of tourism and tourism products in structurally different ways. We shall consider two concepts, local touristic multiplicity and public magnets. Both terms refer to what are called 'pull' factors, i.e. the emphasis is on tourism products and the sub-components of the tourist industry and what can be done on the basis of these to improve the attractiveness of an area. Local touristic multiplicity means:

- A comprehensive range of attractions and activities. These include opportunities for both local and non-local recreation. The conditions here can be divided into place-specific natural and cultural resources and created resources where capital is transformed into tourism products (cf. Svalastog, 1992).
- A developed infrastructure in the area including an effective network of communications both within and to the area.
- A wide range of eating houses and accommodation.
- A comprehensive range of services.

In the development of sustainable tourism the local touristic multiplicity should be directed towards and available to both tourists and residents.

Areas vary in attractiveness. Some areas are so unique that tourists visit them even though little is done to market them. Even if there is a touristic multiplicity, this does not necessarily mean that large numbers of tourists will be automatically attracted. Areas without any special conditions may need a public magnet to provide a viable basis for the tourist industry. Public magnet is the term used for a special activity that will enhance the attractiveness of an area. It may also raise the level of the local multiplicity of tourist activities.

Attractions which are included in the local multiplicity can in many respects be made authentic. They may, for instance, evolve out of the local culture or be based on natural attractions. A public magnet seldom achieves local authenticity. It is generally designed to attract 'mass tourism', to create a demand and later satisfy these needs.

Rooting tourism in the local community can lead to a greater use of local resources in the development of tourism and greater touristic multiplicity through local entrepreneurship. Public magnets are not usually developed on a local basis but on a large scale and rapidly by people who, as regards peripheral areas, are generally not part of the local community. Local touristic multiplicity and public magnets need not necessarily be each other's opposites. Sometimes public magnets are developed and used to 'attract' people to a place; then, by means of various forms of information, these tourists are guided towards the more authentic activities.

What Is Sustainable Tourism Development?

Introduction

> Across the World, the use of the term 'sustainable' has become widespread, and fashionable, so that it is in danger of becoming meaningless. Often, it is not defined, and used in so many different ways that it clearly does not reflect a consensus in meaning'. (p. 4)

As the quotation from Travis (1992) shows, it is essential to define the term 'sustainable tourism development'. We shall attempt this by briefly commenting on the following:

- the concept of development;
- sustainable development;
- sustainable tourism development.

The concept of development

Development has been under discussion for a long time and several schools of thought have been formed. The concept has been dealt with by Friedmann & Weaver (1979), Mabogunje (1980), Stöhr & Taylor (1981), Blomström & Hettne (1981), Hettne (1982), Wahlström (1984) among others. The term is used in many everyday contexts, people speak of the development of an individual, an organisation, a place or society as a whole. What do we mean then by development? The simplest definition is to say that development is change. However, this is not altogether obvious. According to Frängsmyr (in Hettne, 1982) development may have the following meanings:

(1) Change from one state to another, irrespective of direction.
(2) Development from something lower to something higher.
(3) Development from something simple to something more complex.
(4) Development on a linear time scale, i.e. always 'forwards'.
(5) Development 'forwards' also means moral development, towards something 'better' (p. 176).

Development is generally associated with positive social change. However, this is not necessarily the case. Opinions differ, for instance, on the most suitable way of measuring development and on what these measures show. The concept obviously needs to be differentiated. Positive economic development in a society need not mean that developments in the environmental field or within the social system in a society are wholly positive.

It is useful to keep the various possible meanings of the term development in mind as we continue our discussion. It may be of value to ask: development for whom, how and to what?

Sustainable development

The term sustainable development is a recent coinage and has become particularly widespread after the publication of the report 'Our Common Future' by the World Commission on Environment and Development (Hägerhäll, 1988). There are many dimensions to sustainable development and it is by definition normative. This, together with the positive ring to the term, has meant that it has been used and misused in many contexts. 'Our Common Future' has, among other things, the following to say about sustainable development:

- Man himself has the ability to achieve sustainable development.
- A long time-perspective is necessary, resources must last and there must be a good environment for coming generations as well.
- The differences between rich and poor countries must be reduced.
- We must all, and in particular those of us living in the rich world, adopt attitudes and life styles that favour sustainable development.
- Development is a process that can be steered towards sustainability.

There are two fundamentally different approaches to sustainable development. The first is based on the notion that species, ecosystems etc. have a value in themselves and should be allowed to remain in existence on their own terms (Næss in Dahle, 1992). The other takes its point of departure in man and maintains that even if other species have the right to exist, the survival of man comes first.

Sustainable development contains two seemingly paradoxical aspects, namely preservation and development. Thus, it is a matter of simultaneously preserving, for instance, the richness of species and the multiplicity in a natural area, and striving to develop a community in order to achieve a better life for the people.

Sustainable tourism development

Mass tourism

There are many examples around the world of places and areas which have developed mass tourism. The question is whether this form of tourism is sustainable in the long term and, if so, in what fields there is sustainability. Is the economy created by tourism sustainable? Are the changes in social and cultural structures and in the natural environment desirable and acceptable?

One type of study of tourism in poor countries emphasises the fact that most tourism in these countries entails cultural collisions and is a form of neo-colonialism since most of the economic effects of the tourist trade flow back to the rich world via the international companies who own and run the tourist industry. Most of the remaining income goes to the upper class in the poor country (see, e.g. Høivik & Heiberg, 1977, Rossel, 1988, Lea, 1988). This argument would suggest that the inhabitants of the rich world should only be tourists within their own cultural sphere. Is sustainable tourism from rich to poor countries unthinkable?

The tourist's consumption of energy

It is important to distinguish between tourism products where the demand is limited to the local area and products which entail long-distance travel. The problem with long-distance tourist travel is the high level of energy consumption per person it entails. In most cases, high energy consumption leads to high pollution. This problem is taken up by Müller (1992) who has attempted to estimate the consumption of energy per person and kilometre for the various forms of transport used in tourist travel.

To illustrate energy consumption, let us take the example of ecological tourism. What is ecological tourism? Farell & Runyan (1991) explain it in the following terms:

> Cultural ecology concerns human beings and their environment. The relationships of tourists, communities, managers, developers, and policymakers to each others, and especially to their environment, is the substance of ecological tourism and, certainly, sustainable development. (p. 27)

Ecological tourism is a form of small-scale tourism which is based on natural and cultural conditions and does not destroy the resources which tourism builds on. Ecological tourism should also entail development and welfare, e.g. in the form of local economic utility, employment, support for industry and commerce, education, health care etc. in the area the tourist visits (Widstrand, 1993; cf. Laarman & Perdue, 1989; Wilkinson, 1989; Place, 1991; Weaver, 1991; Weiler & Hall, 1992).

Thus, at first sight, ecological tourism would seem to be an excellent form of sustainable tourism. The product is, in many cases, well adapted to both the local community and the environment. However, the initial problem we described, the long-distance travel with high energy consumption which ecological tourism often entails, remains. On the other hand, the high energy consumption per person in ecological tourism must be weighed against the fact that the volume of travel in mass tourism is very much larger and that mass tourism in general entails a high level of exploitation of a place, which often has negative social, cultural and environmental consequences.

Even tourism which is classified as ecological can, if the environment in an area is extremely sensitive, have a serious impact. An example, from tourism in the Antarctic, is given by Hall (1992). Further, ecological tourism may be called into question from the viewpoint that it can lead to a previously unexploited area being opened up for mass tourism.

Tourism's consumption of local resources

> The notion of sustainable tourism has grown out of a response to our overt failure to deal with the random mal-effects of inappropriate tourism development on the environment which hosts it. (Travis, 1992: 1)

There is a considerable amount of literature on the various positive and negative consequences of tourism (see, for instance, Mathieson & Wall, 1982; Murphy, 1985; Pearce, 1981; Gunn, 1988; and Aronsson, 1989). These accounts can be seen as indicators of the impact of sustainable tourism development from a local perspective. At the local level, sustainable tourism development is primarily a matter of preserving the resource basis. Further, tourism should be sustainable in the long-term perspective.

In Sweden natural and cultural resources are crucial to tourism development. However, there is a limit to these resources, which means that they can be overexploited. Like many other industries, the tourist industry has a tendency to build on short-term economic thinking. Unrestrained and unplanned tourism development can result in the deterioration or destruction of the natural and cultural conditions, for instance, disruptions in an area's ecosystem, damage to built resources such as architecture, historical sites or archaeological remains, or the negative impact of tourism development on cultural resources in the sense of traditions, language/dialect etc. (cf. Grahn, 1991). Different communities and areas have different levels of tolerance as regards tourism, depending on how robust they are. The differences may be due to the number and type of visitors, but also to the economic differentiation, social and cultural structure and physical and ecological capacity of the area.

As I mentioned above, sustainable tourism development entails protecting the resource basis but it can also be affected by planning and the choice of sustainable tourism products, i.e. forms of tourism production which adopt a long-term perspective and which make use of (but do not over-consume) resources. A further complement to sustainable tourism development is a combination of appropriate tourism products and selective marketing to different target groups of tourists.

The emphasis on the countryside in Sweden as one of the last wildernesses in Europe is a significant feature in the marketing of Sweden abroad. One of the problems facing the tourist industry is how to earn money from the natural environment, since in our country people have free access to the countryside (allemansrätten — roughly 'right of access'). The answer is simple but, at the same time, difficult to achieve and it would suggest that it is a matter of turning the natural environment into a commodity.

One question that arises is how much can the countryside and nature be commercialised and turned into a commodity. There is probably considerable scope for new forms of nature tourism in Sweden. An important instrument is the possibility of limiting the number of visitors to a place at one and the same time.

An example of one of the first tourism products which has also become one of the primary symbols for Swedish nature tourism is sailing a raft down the river Klar in the county of Värmland. The product idea is simple but ingenious. Somewhat oversimplified, the tourist is provided with timber, rope and a tarpaulin, and there is an instructor on hand if he should run into difficulties. It is then up to the tourist to build his raft and gently sail down the river experiencing nature and tranquillity. The person running the business has imposed a limit on the number of rafts that may be built in one day simply because it is meant to be a wilderness adventure for the tourist and it should not have a negative effect on the environment. The limitations may be compared to the restrictions imposed in some American national parks, where the number of visitors per day is limited.

Many rural areas in Sweden have so far provided very similar products for tourists. The challenge facing the Swedish tourist industry to make it a real source of income in a district is differentiation, developing products with a high degree of uniqueness and finding different competence-enhancing forms of tourism products.

Meeting-places: Categories Based on Supply and Demand

One way of dividing tourism into categories which are based both on the tourists' demands, in the sense of people's values and consumer patterns, and on the nature of the tourism products is the concept of meeting-place. The term refers to the encounter between the tourist and the local environment. The various types of meeting-place are not exclusive categories, dimensions of authenticity can, for instance, be found in several of the categories. The categories are presented below and examples of tourist products in each category are given. Most of the examples are drawn from rural Värmland.

Authentic meeting-places

These are places where the opportunities for encounters are based on local culture and tradition. Places of this type have evolved organically and form a natural part of the total life of the community (cf. Relph, 1976; MacCanell, 1989). However, authenticity is a matter of degree since social change always occurs over time. There is no clear borderline where the genuine becomes the artificial. Encounters in authentic meeting-places generally take place on the residents' terms. The authentic meeting-place can be characterised as a sustainable tourism product in itself as long as it is not changed by tourism. Thus, the critical factor for sustainability is probably the number of visitors. The authentic meeting-place can be overloaded and thus damaged.

An example of this type of meeting-place is the public house in the UK and Ireland. The public house was originally created by the local residents and is still often geared to this group. At the same time, the pub is, in many cases, a tourist attraction, at least for foreign visitors, just because it reveals local colour. An example of a Swedish tourism product which may be considered an authentic meeting place is Mattila Fritid situated in the so-called Finnskogen in the municipality of Torsby in northern Värmland. In Mattila the settlements of the Finnish immigrants are preserved in their original environment. The owner has renovated old buildings authentically; he rents out rooms and cabins, has a restaurant, has prepared paths and running and ski tracks, has a traditional Finnish sauna etc. He sees his facility as part of the development of the district and is also a member of a group concerned with preserving and developing the district as a whole.

Pseudo-authentic meeting-places

The authentic meeting-place is extremely sensitive to exploitation for tourism and to growing demands from tourists. MacCanell (1989) maintains that peripheral areas which were originally authentic but have been 'discovered' and exploited for tourism rapidly lose their authenticity. Thus, one example of the pseudo-authentic meeting-place is an originally authentic meeting-place which has been fundamentally changed but where an attempt is made to maintain superficial authenticity. Another is the museum which in part exhibits authentic objects but where the context is a reconstruction. In this case, then, an effort is made to illustrate authenticity.

An example of a pseudo-authentic meeting-place drawn from Sunne in central Värmland is Martes, a company which has specialised primarily in arranging activity programmes for companies and organisations generally in connection with conferences and courses. These activities may take the form of competitions such as throwing the wellington boot or axe, shooting, and 'put and take' fishing etc. A common feature of most of the activities is that nature and the landscape simply form a background. It should be pointed out that the owners run the company as a form of support fund for local culture and ensure that the activities do not damage the natural setting.

Nature as a meeting-place

Here we are concerned with a limited area which provides the tourist with an experience of nature. The Swedish natural landscape today is largely a cultural landscape in the sense that it has been worked by man for some purpose over a period of time. Nature as a meeting-place may be a sustainable resource and/or product for tourism if it has sufficient carrying capacity or if visitors are channelled into suitable areas. Thus, the same restrictions are valid here as for authentic meeting-places.

An example of nature as a meeting-place is walking as a tourist in a national park or nature reserve with paths and rest cabins which offer encounters involving an experience of nature. A concrete example is Hovfjället nature reserve in the municipality of Torsby in northern Värmland. The park has facilities for downhill and cross-country skiing and for walking and a restaurant/café. One of the activities offered is guided nature tours. Adjoining the park there are two holiday villages.

Instrumental meeting-places

These refer to opportunities for meetings which are generally arranged locally. The tourist and the resident have a common aim in the encounter. It is typical of instrumental meeting-places that a specific encounter occurs during a short period. Examples of this type of meeting-place are provided by various types of events such as runs and festivals, or sites such congress halls. Even in the case of instrumental meeting-places the extent of the tourism and the type of activity are decisive for sustainability.

An example of an instrumental meeting-place is the folk musicians' rally in Ransäter in the municipality of Munkfors in central Värmland. This rally takes place in June and attracts tens of thousands of people.

Commercial meeting-places

These are artificial places constructed with the aim of creating and satisfying socially generated commercial needs. They are primarily designed to entice long-and short-distance visitors or tourists. They have seldom evolved organically but have been built up relatively rapidly along functional lines for commercial purposes. The commercial meeting-places are often simplified, standardised pseudo-places which lack an original local identity. Examples are theme parks of the summer land type or whole cities like Las Vegas.

An example of a commercial meeting-place in Sweden is Branäs, a large recreation centre which has necessitated major financial investment in the municipality of Torsby in northern Värmland. Branäs is primarily a downhill skiing facility in winter but in the summer it offers a whole range of activities. Rooms and cabins are available and there are restaurants etc. The facility may be seen as a public magnet for northern Värmland.

In most cases, commercial meeting-places do not result in sustainable tourism development from an ecological and cultural perspective. There are several reasons for this. Firstly, they are, of course, part of the economic market where the keynote is commercialism. This life style hardly leads to a good use of resources in a global perspective. Secondly, these meeting-places are intended to generate a considerable amount of travel with probable pollution effects. Thirdly, in most cases, the actors developing the commercial meeting-places do not heed the original local culture.

The exclusive meeting-place

This is a special form of the commercial meeting-place. This type of meeting-place is basically commercial, but it is for the relatively few who can afford it. It is more a matter of being seen (vanity) than of being. This meeting-place can be consider an expression of the post-modernist society. The tourism product can be sustainable in certain respects, for instance it generates relatively few visitors, but at the same time it is artificial and generally does not heed the original local culture. An example is the exclusive casino. A Swedish example of an exclusive meeting-place is the Selma Lagerlöf Hotel and Spa situated in Sunne in central Värmland. The Spa facility offers health care and exclusive service, with medical staff available.

If the tourists become dominant at any of these meeting-places, this may lead to conflicts with the local residents. By their numbers the tourists as a collective can dominate a place or control a time-space. The greatest risk of disruption and conflict is probably at the authentic meeting-place or in nature as a meeting-place. The purpose of instrumental and commercial meeting-places is, of course, to attract large numbers of people.

Concluding Remarks

Sustainable tourist development must be based on the whole tourism system, i.e. the tourists' demands, the travel itself and the meeting-places between the tourist and the local environment with its offering.

As regards demand, sustainable development is a matter of changes in attitudes and greater knowledge in the individual, for instance, taking active responsibility for the environment by reducing energy consumption, buying recyclable goods and dealing with waste. At the same time, steering demand is a matter of changing national and international regulations and agreements to favour long-term sustainable development. A change towards a more sustainable society naturally requires some agreement among governments and other actors as to the criteria for sustainability.

Sustainable development has several dimensions. One concerns economy where developments may move from local to international cycles, a trend which has characterised production and ownership in Swedish industry during the last century (Melin *et al.*, 1984). We have pointed out earlier that sustainable tourism development should lead to good income opportunities for the local community. This causes us to emphasise the importance of the local and regional economy.

A second dimension concerns ecology where the trend may be from a destructive to a maintaining impact. Sustainable tourism development means limiting consumption of and damage to both natural and cultural resources to a minimum; any 'development profit' should, if necessary, preferably be used to improve worn-down resources and environments.

The third dimension covers social and cultural values which may vary on a scale from preservation to change. The two concepts of preservation and change in the form of development are often recurrent in the discussion of sustainable development. There is a tendency towards value conservatism if 'everything' is to be preserved in accordance with tradition. Everything that is traditional in a community need not be good. The paradox for sustainable development is that there must be some change in a place. The question is who is to decide what is a good change. The answer, from my standpoint, is that the extent and rapidity of change is best determined by means of the democratic process in a society if this is well-developed and efficient. This reasoning implies that people at the local level should have major democratic influence over tourism development since it is there that the impact is most evident. Even this point of departure may be problematic. An example is to be found in the situation where the people in a particular place are faced with the choice between an increase in the number of jobs from a new tourist facility and the fact that the facility entails an encroachment on the natural environment. What is the choice? Particularly when it comes to natural resources and nature conservation, a holistic assessment is essential.

An alternative point of departure is to find a catalyst in a community, e.g. a competent tourism planner who can both provide the necessary 'input' of novelty and knowledge etc. and plan for the preservation of fundamental and 'good' traditions.

To summarise, conditions of the following type are generally emphasised in definitions of sustainable tourism development:

- Sustainable tourism development should entail a long-term perspective for the use of resources in tourism production.
- A form of tourism should be developed that contributes to creating equality and economic and social welfare for the local community.
- Tourism development should make careful use of the natural and cultural conditions, including the built environment.
- Positive tourism development in line with the above points should not be a burden on other people and areas or on coming generations.

Finally, I would like to outline some principles which lead to sustainable tourism systems.

(1) Before undertaking a trip the tourist should read and learn about the culture, nature and society at his destination. This can be achieved by means of study-circles arranged by the travel agent, who also provides comprehensive material.
(2) Increase the knowledge and formal competence in a community on issues concerning sustainable tourism, particularly as regards product development and planning.
(3) Development opportunities for recreation in places, in particular in the vicinity of big cities. This may help to eliminate part of the long distance and energy-demanding travel from the cities.
(4) In peripheral areas, invest in a greater number of products within authentic, certain types of instrumental and pseudo-authentic meeting-places and nature as a meeting-place. Although these involve long-distance journeys, they will entail a relatively small degree of exploitation if limits are set from the start for the number of visitors at one and the same time. The meeting-places mentioned generally have good local economic utility.
(5) If commercial meeting-places with the character of public magnets are to be developed, site them mainly in densely-populated areas which are better able to cope with the great pressure of visitors than the periphery is.
(6) In peripheral areas, integrate tourism both economically and socially in the local community. This can be achieved by developing slowly and on a relatively small scale in order to provide the best opportunity for the emergence of economic networks between tourism and other industries. Furthermore, use resident labour in the tourism industry, alternatively create opportunities for newly arrived tourism labour to become resident.
(7) Support local initiatives and give people the opportunity to participate in the planning process in tourism development. The planning of tourism development should adopt a holistic perspective. The whole of the tourism system should be taken into account and tourism should be placed in the context of other industries and other development measures.

References

Aronsson, L. (1989) *Turism och lokal utveckling. En turism-geografisk studie (Tourism and Local Development. A Study in Tourism Geography)*. Reports from the Departments of Geography. Series B. No 79. (Dissertation). Department of Human Geography, Gothenburg University Business School.

— (1993) *Mötet. En studie i Smögen av turisters, fritidsboendes och bofastas användning av tid och rum (The Encounter. A Study of How Tourists, Vacation and Permanent Residents in Smögen Utilize Time and Space)*. Regional Science Research Unit. University of Karlstad. Research Report 1993: 1.

Blomström, M. and Hettne, B. (1981) *Beroende och underutveckling. Den latinamerikanska beroendeskolans bidrag till utvecklingsteorin (Dependence and Underdevelopment. The Contribution of the Latin American Dependence School to Development Theory)*. Prisma.

Clawson, M. and Knetsch, J. (1966) *The Economics of Outdoor Recreation*. Baltimore: Johns Hopkins University Press.

Cohen, E. (1974) Who is a tourist? *The Sociological Review* 22 (4).

Farrell, B. and Runyan, D. (1991) Ecology and tourism. *Annals of Tourism Research* 18 (1). (Special Issue: Tourism Social Science).

Friedmann, J. and Weaver, C. (1979) *Territory and Function. The Evolution of Regional Planning*. London: Edward Arnold.

Frängsmyr, T. (1982) Commentary on Björn Hettne's Report: *Strömfåra och kontrapunkt i västerländsk utvecklingsdebatt (Mainstream and Counterpoint in the Western Debate on Development)*.

Grahn, P. (1991) Using tourism to protect existing culture: A project in Swedish Lapland. *Leisure Studies* 10 (1). Sid 33–48.

Gunn, C. A. (1988) *Tourism Planning* (2nd edn). Basingstoke: Taylor & Francis.

Hall, M. (1992) Tourism in Antarctica: Activities, impacts, and management. *Journal of Travel Research* No 4.

Hettne, B. (1982) *Strömfåra och kontrapunkt i västerländsk utvecklingsdebatt (Mainstream and Counterpoint in the Western Debate on Development)*. Background Report 8. Naturresurs- och miljökommittén. Stockholm.

Hägerhäll, B. (ed.) (1988) *Vår gemensamma framtid. Rapport från världskommisionen för Miljö och utveckling under ordförandeskap av statsminister Gro Harlem Brundtland (Our Common Future. Report from the World Commission for Environment and Development under the Chairmanship of Prime Minister Gro Harlem Brundtland)*. Stockholm: Bokförlaget Prisma och Tidens förlag.

Høivik, T. and Heiberg, T. (1977) *Tourism, Selfreliance and Structural Violence*. International Peace Research Institute. PRIO. Publ. S-14-77. Oslo.

Karlsson, S-E. (1994) *Natur och kultur som turistiska produkter — en börjanfill en sociologisk analys (Nature and culture as touristic products — A tentative sociological analysis)*. Research report, Regional Science Research Unit, University of Karlstad.

Krippendorf, J. (1989) *The Holiday Makers. Understanding the Impact of Leisure and Travel*. Oxford: Heinemann Professional Publishing.

Laarman, J. and Perdue, R. (1989) Science tourism in Costa Rica. *Annals of Tourism Research* 16 (2), 205–15.

Lea, J. (1988) *Tourism and Development in the Third World*. London: Routledge.

Lowyck, E., Van Langenhove, L. and Bollaert, L. (1992) Typologies of tourist roles. In P. Johnsson and B. Thomas (eds) *Choice and Demand in Tourism* (pp. 13–32). London: Mansell.

Mabogunje, A. L. (1980) *The Development Process. A Spatial Perspective*. London: Hutchinson.

MacCanell, D. (1989) *The Tourist: A New Theory of the Leisure Class*. London: Macmillan.

Mathieson, A. and Wall, G. (1982) *Tourism: Economic, Physical and Social Impacts*. London and New York: Longman.

Melin, L., Alsén, G., Berger, S. and Nilsson, A. (1984) *Kommunerna och näringslivet. Ett näringslivspolitiskt dilemma (Municipalities, Industry and Commerce. An Industrial Policy Dilemma).* Stockholm: Publica.

Müller H. (1992) Ecological product declaration rather than 'green' symbol schemes. *The Tourist Review* No. 3.

Murphy, P. E. (1985) *Tourism: A Community Approach.* New York: Methuen.

Næss, A. (1992) Bærekraftig utvikling: En begrepsavklaring (Sustainable Development. Definition of Concepts). In B. Dahle (ed.) *Bærekraften for næringsutvikling i bygdesamfunnet: Holdninger, Kultur, Miljo, Natur (The Viability for Commercial Development of the Rural Community: Attitudes, Culture, Environment, Nature).* Norwegian College of Physical Education. No. 107.

Ogilvie, F. W. (1933) *The Tourist Movement.* London.

Pearce, D. (1981) *Tourist Development. Topics in Applied Geography.* Harlow: Longman.

Place, S. (1991) Nature tourism and rural development in Tortuguero. *Annals of Tourism Research* 18 (2), 186–201.

Relph, E. (1976) *Place and Placelessness.* London: Pion.

Rossel, P. (1988) Tourism and cultural minorities: Double marginalisation and survival strategies. In P. Rossel (ed.) *Tourism: Manufacturing the Exotic* (Document 61, pp. 1–20). Copenhagen: International Work Group for Indigenous Affairs.

Stöhr, W. B. and Taylor D. R. (eds) (1981) *Development from Above or Below? The Dialectics of Regional Planning in Developing Countries.* New York: Wiley.

Svalastog, S. (ed.) (1985) *Turisme ved kysten. Inpassing av reiseliv i lokale samfunn (Tourism on the Coast. Adapting Tourism to the Local Community).* Information Series No 55. Oppland District College, Lillehammer.

— (1992) *Hva har skedd i Norsk reiseliv i de siste 50 år? (What has Happened in Norwegian Tourism During the Last 50 Years?).* Paper presented at the UNESCO conference on Røros.

— (1993) *Lokalisering av reiseliv. Om resursanlyser, den romlige fördelning og lokal innpassing. (The Location of Travel. On the Analysis of Resources, Spatial Distribution and Local Adaptation).* The Norwegian School of Business Administration and Economics, Department of Geography, Bergen.

Sveriges Turistråd (1986) *Fakta om fritid. Några aktuella undersökningsresultat (Facts about Leisure. Some Current Research Results).* No. 4. Stockholm.

— (1991a) *Sverige — något att satsa på. En skrift från Sveriges Turistråd om turismens betydelse (Sweden — Something to Invest In. A Pamphlet from the Swedish Tourist Board on the Significance of Tourism).* Stockholm.

— (1991b) *Svenskarnas resande i Sverige. Regional omfattning och betydelse. (Swedes' Travel in Sweden. Regional Extent and Significance).* Forskning och utveckling (Research and Development). No. 9. Stockholm.

Travis, A. S. (1992) Sustainable tourism planning and development. A paper for Norway and Sweden. Section for Geography and Tourism, Dept of Social Sciences, University of Karlstad.

Wahlström, L. (1984) *Geografiutveckling och geografisk utveckling...som om platser betydde något (The Development of Geography and Geographical Development...as if Places Meant Anything).* Department of Human Geography. Gothenburg University. Series B, No. 76.

Weaver, D. (1991) Alternative to mass tourism in Dominica. *Annals of Tourism Research* 18 (3), Sid 414–432.

Weiler, B. and Hall, M. (eds) (1992) *Special Interest Tourism.* Belhaven Press/Halsted Press.

Widstrand, S. (1993) Vad är Ekoturism? (What is Ecological Tourism?) Travel Magazine *Vagabond.* Appendix No. 3, pp. 40–42. Stockholm.

Wilkinson, P. (1989) Strategies for tourism in island microstates. *Annals of Tourism Research* 16 (2), 153–77.

Zimmermann E. W. (1972) *World Resources and Industries* (3rd edn). New York: W.N. Peach. James A. Constantin. Harper and Row Publishers.

Home, Escape and Identity: Rural Cultures and Sustainable Tourism

David Crouch
Department of the Built Environment, Anglia University, Chelmsford, Essex, CM1 1LL, England

Tourism is an increasingly significant part of contemporary culture. This paper weaves new ideas in the theoretical interpretation of tourism, culture, and the environment together with a discussion of how sustainable tourism might be distinguished from other forms of tourism in rural areas. It examines the ways in which tourism development engages with the people who live and work in the countryside and their heritage. Finally, initial suggestions are made about a new approach, which could be used in Western Ireland. Informed theory is used to understand observed reality to provide a base for practical application.

The Cultural Situation Surrounding Tourism in Rural Britain and Ireland

Our contemporary period has been interpreted as *late modernity*, or *post modernity* (Urry, 1990). Under this labelling, there are important components which have a profound impact on the way people seek to use rural areas for tourism, and on how entrepreneurs, local councils, central governments and others may seek to develop tourism.

The cultural shift towards *post modernity* is characterised by many factors — by a sense of rootlessness, the importance of the visual in the tourist experience, the significance of detachment, of extreme sensations and fleeting movements through space, of experiences which are separated from everyday experiences, which exist outside our 'known communities' and places (Urry, 1990). The allure of authenticity — something that is demonstrably grounded in history and is celebrated as such — coexists with a desire for the non-authentic — something obviously 'plastic', that is enjoyed for its own sake and is not rooted in other associations and pasts (Wright, 1985; Hewison, 1987). Both 'the environment' and 'people as culture' become important in this process (Urry, 1992). The position of the environment, including rural space and landscape, is especially important here, both as a feature in its own right and as some kind of backcloth. Cultural activities are often seen as growing from the environment. This highlights the action of 'the gaze' — of looking in a detached way at events and excitements — and the desire for something quite different from everyday knowledge and experience, as well as the increasing fascination for discovering roots, a fascination which has helped heritage to be a particularly useful tool for the promotion of tourism (Wright, 1985; Hewison, 1987). Heritage, paradoxically, has become effective both as historically rooted heritage, and as 'manufactured' heritage. In both cases, the central position of heritage relates to the popular desire to experience, perhaps fleetingly, a phenomenon so distinctly and obviously unconnected with the everyday, that it becomes reminiscent of journeys abroad to so-called native cultures, an experience of the *Other*, of worlds apart (Cohen & Taylor, 1992/1978). And, at the same time as this shift in popular culture is taking place, the increasing process of commercialisation is placing new demands upon cultural change, and on the lives of ordinary people (MacCannell, 1992). It is also throwing up a series of new opportunities for the tourist and for the tourism developer.

This complex process can be related to wider reinterpretations of rurality and the countryside in western society (Crouch, 1992a). Pertinent components of this reinterpretation include the replacement of the idea of rural as being community and continuity led by a demand for diversity: rurality appears less as a unified alternative to urbanity than simply another source of variety, opportunity and enterprise. This has been appreciated in the UK, in Ireland and elsewhere with the erosion of the primacy of agriculture in the image of the countryside, challenging the familiar dominant organisations of farming and conservation. There is a restructuring of the rural political economy and a drive towards new forms of capitalisation, alternative sources of profitability, market opportunism and to the assertion of consumer choice. There is a mixing of heritage attractions with new attractions, with golf being an archetypal case. Amidst this wave of change there is the fascinatingly paradoxical borrowing of features of imagined rural Britain, France, America and Ireland, as imaginative wrappings used to re-present new rural enterprises, including tourism enterprises (Crouch, 1992a).

The importance of national identity in this process is profound. For Ireland , rural areas form a central component of the national identity (Newman, 1967; O'Connor, 1993). In turn, the role of rural Ireland's heritage has been pivotal in the promotion of Ireland as a tourism destination. That promotion has become a particular abstraction of culture (Sheeran, 1988a, 1988b). It is connected with a classical concern for the changing rural landscape; its converse is expressed in the rash of modern bungalows spreading through the traditional Irish countryside. This may, however, reflect a lack of concern about architectural attributes amongst contemporary rural Irish people rather than any dilemma over design and style (Sheeran, 1988a). Tourism's handling of Irish heritage as a promotional tool may equally be the replacement of an older culture with a new tourist-orientated culture, and, moreover, it may be intended that the new culture will *last*. Whether this makes traditional culture especially vulnerable is explored later in this paper.

There is a curious irony in the position of heritage *vis à vis* tourism; that position is especially well illustrated in Ireland's case. Ireland is seen to offer an almost perfect *Other* for tourist fascination (O'Flaherty's famous 1930s film 'Man of Arran' is symbolic of that wider process). However, tourism to Ireland also offers *Home* (Crouch, 1992b). The way in which *home* is promoted may be compromised in the popular styles of tourism presentation and promotion. There is, for example, a mechanistic approach to tourism policy in Ireland, seeking quick returns on investment, in terms of cash flow, and jobs created. This is surely to be expected, but deeper questions deserve to be raised, especially about the dilemma of the meaning of *sustainability*.

Caricatures have become a crucial part of Ireland's tourist promotions, presenting a land of peasants and of a disabled economy, and the various social and economic consequences which follow. But change these caricatures and have you lost Irish tourism? Do they allow for change? Is any other culture redundant if it cannot be related to tourism by being converted to caricature? Do the cultural features used by tourism acquire a different quality through this process? Culture is itself a problematic word, although tourism promotion often uses it as a taken-for-granted category. This may be important in terms of the way sustainability is understood and connected policy objectives are forged.

Of course it has been argued powerfully in countries that sustain significant cultures which are distinct from the major West European and American ones that there is both the possibility of using heritage in promotion and in attractions, and at the same time experiencing its erosion by tourism. These are real possibilities, and there are quantitative survey methods to investigate these processes. This paper explores the cultural context within which we may alert policy makers to the predicament of traditional cultures, how we can further develop our understanding of sustainability in that process, and it points towards means of bringing quantitative research approaches to bear on the issues involved.

There are many forms of culture, and of cultural identity. Tourism promotion uses ideas of cultural identity with which visitors can identify by association. Identity is, however, complex and changing. For example, there are locals whose culture is often abstracted in promotion. But there are also new people moving in to areas, there are others purchasing second homes there, and yet others who return to an area after a period spent living and working elsewhere. All this produces a complex and dynamic cultural picture, although investigations in Britain have frequently reduced this complexity to insider–outsider polarities, and have denied the continuing process of change that can be mutual, so that the distinctiveness is sustained, but the differences dissolve. This may be a process less of erosion than of transition, a dynamic process. What matters is how the relations of power operate (Byrne, Edmondson & Fahy, 1993).

Heritages, Environments and Identities

Implied rural 'local' cultures of the type discussed above are often seen as being related to the specific environments in which they happen, and this is how tourism presents and represents these phenomena. Tourism development may, however, change both rural cultures and rural environments. What is tourism's position about these implications? There are important questions of identity here: whose identity is being presented to visitors, and for whom is that presentation being made? Marketing materials may offer escape from one culture and identity into another. Visitors may envisage making that escape, but how will place and culture be related in those materials? Questions such as these appear to be fundamental to heritage presentation for both individual enterprises and for public policy.

The heritage picture which is presented to the visitor may be quite different to that experienced daily by those who live in the countryside. This will be a complex picture: rural populations are nowadays multi-faceted, even to some extent multi-cultural (see Shurmer-Smith, 1990 on the French experience; Urry, 1992). Mormont (1987) has explored such issues in the Ardennes. There a complex series of alliances was revealed crossing over local and less local people, and operating around questions of commercialism, of scale, and of conservation. Mormont focuses on the issue of control: how much local people feel that they can have an influence, and a sustainable influence, over tourism development. This points towards issues of both sustainabilty and local empowerment

There are important political questions here about power over space and environment. The questions are two dimensional. There is the real and immediate 'concrete' reality of territory and space. There is also an underlying sense of identity with that space which is a symbolic and cultural phenomenon. In the Irish context the icons and the complexity of this two-dimensional reality were dramatically shown in Jim Sheridan's film 'The Field'. In that film, the land was of central importance, first as a source of livelihood, but equally in a very powerful historical context, as a representation of a sense of struggle. The landscape as environment, the grand, uneasy and beautiful setting in which it happened, was a visible representation of that heritage, without which the story would be very different. And there is a crucial (perhaps cruel?) story there about how much the significance of history can be contained within any new cultural presentation (Wright, 1985; Hewison, 1987)

In 'The Field', the American arriver, Berringer, comes to make profit to take away. There is a different kind of American in Robert Bernen's *Tales from the Blue Stacks* (1978). Bernen lived and worked on a piece of land in Donegal, partly as observer, eventually writing a book on the landscape, life and people of the place, but also working the land. This experience was rather different from 'refurbishing cottages'. What kind of distinctions can be drawn between this presentation of heritage and the more commonly available one presented to tourists? This is not a simple question of authenticity, but one about relationships, distinguishing perhaps between the exploitative and the sustainable.

Central to this question is the debate about how rural environments and cultures are consumed. The idea of visitors making a relationship with people who live in the countryside — farmers, musicians, shopkeepers etc. — may seem both absurd and nostalgic. Yet there is evidence that part of the sought after escape is an escape *into* a culture which may seem exotic to the outsider, but which is another's everyday life. Research and writing in the UK, in France and in the United States points to different possibilities in the ways in which leisure is realised. Even the original Disneyland was based on an idea for a happy settlement for people of all kinds to live in. Tourists to France seek out the 'homely, rustic cuisine' (Zukin, 1991). People get away to second homes in log cabins to celebrate 'family', and get together in 'street communities' after a drive through countryside in their caravans (Wolfe, 1977). Escape becomes an escape for home, not just from home. There is an often confusing mixture of homeliness and reassurance, of surprise and difference in leisure (Crouch, 1992b). It may be, however, less that the market responds to demand than that it reacts to what is profitable and then seeks to influence consumers along those lines. Does the promoted friendliness of Irish culture become particularly important in this search? The faces in the brochures welcome us to a happy culture — the face as living heritage — but the opportunities to take that forward into real engagement may present more challenges.

These issues point towards questions of contemporary cultural consumption. How do visitors want to consume particular cultures and environments? How do cultures and environments and their peoples want to exploit that consumption, if at all? This is a complex interaction, made difficult by the many different points of reference and cultural identities in Western Europe (Massey, 1993). As a result, the ways in which areas are experienced are culturally contextualised, and there is a feedback into what they do to and with those areas. Further, like clothes and making music, places are given particular cultural value through the way they are used in everyday life, in addition to their role as icons of place, history and culture. For many peripheral areas, such as rural Ireland, generations of negative imagery make a transformation into positive imagery difficult.

A further point to be noted here is the concept that different cultural values may be given different spaces within the menu presented for the visitor. There can be front spaces for presentation and celebration, and back spaces for privacy and withdrawal. There are different settings for types of social intercourse, and different kinds of cultural celebration and knowing (Goffman, 1959; Giddens, 1984; Urry, 1992). Failure to understand these hierarchies can lead to confusion and conflict. These conflicts can occur at the local level; they can also occur at the interface between nationally promoted identities and locally perceived identities.

How can heritage be reworked to avoid conflict? How can heritage become more than features on a list that may be used to attract people, simply as objects of tourist *gaze*? How can heritage become involved with the complex motives of the visitor discussed above?

The Possibilities in Community Based and Sustainable Tourism

A sustáinable approach to tourism envisages the people of an area being empowered by owning their own culture as part of any tourism development. Rural tourism may thus seek to replace or supplement abstracted heritage with a different set of values and experiences. It may begin to see the artefacts which have been retained in reserves or at visitor centres as curiosities in a wider range of heritage experiences.

In recent papers Urry has confirmed that possibility, by refining his *gaze* to present several categories: romantic, collective, spectatorial, environmental and anthropological. Four of these (omitting spectatorial) involve either or both sustainability and shared encounters with local people. Perhaps the commercial approach has got it wrong ! This again points to power, and it may be that the emphasis of tourism in the 1980s was simply a product of the commercial political dynamic of the time. Future trends will, however, be closely influenced by the policy and practice of those agencies and individuals involved in tourism development.

Process and Practice in Developing Sustainable and Community Tourism

The central task is that of constructing cultural identities in true participation with the community, identities which are often very different from those sought by a mainstream commercial approach. Although these identities may not be based on a financial/economic relationship there is no reason, however, why they should not be successful financially. But they involve different kinds of presentation, activities, practices and forms of engagement that are often unashamedly social and cultural. Different self-images are generated amongst hosts and visitors alike. The key implication in this approach is that of confidence building as part of tourism management: central to this confidence is the process of empowerment for local communities Without this reorientation, sustainable tourism simply recycles old methods such as 'participation' in a new language. What are considered 'proper' management and presentation skills may mean that tour officials and advisers impose patterns and procedures that ignore the patterns of specific local cultures involved and at the same time can diminish the enjoyment that the visitor seeks. This texture of visitor experience may not emerge under the standard headings of visitor questionnaires, nor through the typical ways in which tour managers converse with local people (Embacher, 1994). Such questionnaires may undermine the relationship based upon local cultures which actually attracts people to an area. Father McDyer's pioneering work in the 1970s in Glencolumbchill in County Donegal is informative in the ways in which the community wove its own cultural interpretation methods for the then newly emergent local visitor industry (Keane & Quinn, 1991). Although these experiments were financially unsuccessful, their failure did not arise because of the visitor experience techniques used: it should be noted that many major theme parks have also had their financial difficulties. Moreover, the new art that is still produced in Glencolumbchill has made new icons for contemporary Ireland.

Writing on Ireland, Patrick Sheeran (1988b) is only half right when he says, perceptively, 'it is pertinent to enquire why anybody should want to visit us...why spend time in the Irish countryside, one of the most unkempt and certainly most littered in Europe'. They spend time in Ireland, because, significantly, it is a wonderful *Other*, and has much for the postmodern imagination too. But this is not to argue for conserved poverty, nor for the poverty of the surviving culture through a separate commercial success.

Any project for dynamic cultural development should seek to engage the everyday lived and changing culture of the local people who remain when the visitors have left, as well as the preferences of the visitor. This engagement between their several interests should be achieved in a sustainable and not exploitative way. Visitors will make their own icons — photographs, sketches, postcards, even memories — and then take them home. Tourism may therefore reconstruct place and culture, but those who live locally can still recontextualise the imposed changes to make them theirs. And care may be necessary when we use words like 'inappropriate development' — there may be mutual interest if it also engages the lives of the people who spend their lives there, and be very 'appropriate' (Gilbert, 1993).

However, as Lengeek (1993) remarks, 'in the recent past critical consumer behaviour (not simply from consumers tired of a hackneyed presentation) has not been encouraged. For this to happen, organised interest groups and government need to create visions of the development of our social space'. Central to the approach to sustainability is a shift from consumerism to culture; from consumer to citizen (Zukin, 1991; Crouch, 1992b, 1994). Rural tourism can be cultural empowerment.

The Welsh commentator and consultant Arwel Jones (1993) has referred to the need to draw attention to customs, culture and language in order to promote an understanding and respect for the tourism resource base of the natural, built and socio-cultural realms. He advocates working with local communities to secure appropriate action, forming partnerships, the dissemination of good sustainable practice, and monitoring and review processes (Heneghan, 1993). On a different level these ideas could be linked to procedures for granting planning permission, for developing tourism promotion and ultimately to the time horizons of the policies of the European Union, with this last European level having an over-arching potential to enable or disable (however well meant). For example, short-term success may trivialise the process of sustainability. Simple economic measures, with their familiar short time horizons, may be unhelpful. Investment programmes for welfare and for economic development (and the positioning of tourism in these programmes) may cancel one another out, and need therefore to be targeted differently. Certainly, restrictions on the earning capacity of those on welfare payments in the western parts of Ireland have had a negative effect on small-scale tourism development. There are numerous other practical steps which could directly respond to this kind of reasoning. Practical steps could include:

- Clarity of purpose in public policy, including relating tourism to employment, welfare and the local economy.
- Environmental and heritage policies which are geared to living heritage and not simply the past.
- An appreciation of the multiplier effects of local tourism, and a wider definition of profitability.
- The use of partnership and enabling techniques, now being pioneered by some Leader Projects.
- The recognition of the importance of empowerment.
- Recognition of the need to bring in professional marketing and financial expertise.
- Giving prominence to local cultural expertise, experience and ideas, once released from generations of negative imagery. Empowerment to create without copying international theme park styles.
- Initiatives at all levels and in many forms.
- Interpretation centres that are designed as multi-purpose centres for local as well as visitor use.

These kinds of argument challenge the lazy approach of 'giving the customers what they want'. The customer's perception of place, its facilities and its culture is shaped in part by promotion. The 'want' is structured and so open to being changed. There may be tourists eager to escape their world, but there is still greater scope for sustainable tourism that responds to the search for *home*. Standard market research, with its tight packages of topics and merely quantitative information ignores many of the questions raised in this paper, questions which necessitate a difficult process of lateral thinking.

Research Footnote: Finding a Base for Sustainable Tourism

There is an important practical project to unravel and reconstruct culture and landscape, history and lived culture, and to explore their appeals to both tourism and continuing cultural development. This empirical investigation could benefit both the debate about the *gaze* and tourism development generally. Such a research process would be useful for practical policy makers if it used ethnographic investigation, including the visual ethnography of environment and space in the everyday lives and cultures of hosts and visitors alike.

The Gaeltacht of Ireland would provide an ideal focus for an investigation of this kind: that investigation could provide answers for many other peripheral cultures central to the tourist trade, as well as to European Union social policy. The Gaeltacht comprises a series of areas, largely in the west of Ireland, where the Irish language has survived in daily use. Approximately 55,000 people live in the Gaeltacht areas, where special measures are employed to assist the progress of Irish. The Gaeltacht is, however, not just a linguistic concept: it possesses distinct forms of local social relationships within a context of the continuity of a traditional culture. These areas are a great attraction for visitors, who make both temporary and repeated trips.

The investigation could explore a variety of topics. Do visitors and hosts have shared values? What is their involvement together? Do visitors return because it is *Home* to them? Are there differences between those with researchable forebears there, and those without? How does experience relate to promotion? Does cultural change relate to both old stable and declining cultures and the new/postmodern? The processes involved in this investigation would be numerous but would involve in depth interviews with both visitors and locals, and with tourism 'professionals' and managers.

References

Bernen, R. (1978) *Tales from the Blue Stacks.* London: Hamish Hamilton.

Byrne, A., Edmondson, R. and Fahy, K. (1993) Rural tourism and cultural identity in the west of Ireland. In B. O'Connor and M. Cronin (eds) *Tourism in Ireland: Critical Perspectives.* Cork: Cork University Press.

Cohen, S. and Taylor, L. (1992/1978) *Escape Attempts* (2nd edn). London: Routledge.

Crouch, D. (1992a) Popular culture and what we make of the rural. *Journal of Rural Studies* 8.

— (1992b) Popular culture and the distinctive forms of leisure cultures: Leisure as homes in other places. *Proceedings of the European Leisure Research Association Conference, Bilbao.*

— (1994) Tourism and the back garden: Questions of power and space in self generated leisure. Working Paper. University of Warwick (in press).

Embacher, H. (1994) Marketing for agri-tourism in Austria: A strategy for a highly developed tourist destination. *Journal of Sustainable Tourism* 2 (1&2).

Giddens, A. (1984) *The Constitution of Society* (pp. 124 *et seq.*). Cambridge: Polity Press.

Gilbert, D. (1993) Issues of appropriate rural tourism development in southern Ireland. *Leisure Studies* 12, 137–46.

Goffman, E. (1959) *The Presentation of Self in Everyday Life.* Doubleday.

Heneghan, M. (1993) Co-operative rural development. Paper given to the conference on Co-operative Rural Development held in Westport, Co. Mayo, Ireland.

Hewison R. (1987) *The Heritage Industry.* London: Methuen.

Jones, A. (1993) Sustainability and community participation in rural tourism. *Leisure Studies* 12, 147–8.

Keane, M. and Quinn, J. (1991) Community tourism in rural Ireland. In M. Cuddy *Rural Development in Ireland.* University College Galway.

Lengeek, J. (1993) Collective and private interests in recreation and tourism — the Dutch case: Concerning consequences of a shift from citizen role to consumer role. *Leisure Studies* 12, 7–23.

MacCannell D. (1992) *Empty Meeting Grounds.* London: Routledge.

Mormont, M. (1987) Tourism and rural change: The symbolic impact. In M. Bouquet and M. Winter (eds) *From Whom Their Labours Rest: Conflict and Practice in Rural Tourism.* Chichester: Avebury.

Massey, D. (1993) Power geometry and a progressive sense of place. In Jon Bird *et al.* (eds) *Mapping the Futures.* London: Routledge (see also papers by Jackson, Harvey and Puttnam in same publication).

Newman, J. (1967) *New Dimensions in Regional Planning.* Dublin: An Foras Forbartha.

O'Connor, B. (1993) Myths and mirrors: Tourist images and national identity. In B. O'Connor and M. Cronin (eds) *Tourism in Ireland: Critical Perspectives.* Cork: Cork University Press.

Sheeran, P. (1988a) Genius fabulae: The Irish sense of place. *Irish University Quarterly* 18, 191–206.

— (1988b) The idiocy of Irish rural life reviewed. *The Irish Review* 5, 27–33.

Shurmer-Smith, P. (1990) Home is where the heart is: Second home ownership on Ile-aux-Moines, South Brittany. Mimeo quoted in Urry (1992).

Urry, J. (1990) *The Tourist Gaze.* London: Sage.

— (1992) The tourist gaze and the environment. *Theory, Culture and Society* 9, 1–26.

Wolfe, R.I. (1977) Summer cottages in Ontario: Purpose-built for an essential purpose. In J. Coppock (ed.) *Second Homes: Curse or Blessing.* Oxford: Pergamon.

Wright, P. (1985) *On Living in an Old Country.* London: Verso.

Zukin, S. (1991) *Landscapes of Power.* University of California Press.

Sustainable Rural Tourism Strategies: A Tool for Development and Conservation

Bernard Lane
Rural Tourism Unit, Department for Continuing Education, University of Bristol, Bristol BS8 1HH, UK

This paper reviews how the principles of sustainable tourism have special relevance to the development of rural tourism, and examines how those principles can be translated into practice by the writing and implementing of regional sustainable tourism strategies. It considers the advantages of this approach, and offers guide-lines for future practitioners. A case study is given of the development of a strategy for an area in northern England, Berwick-upon-Tweed.

> *Travel and Tourism is the world's largest industry and the major contributor to global economic development. One in 15 employees — 112 million world wide — is involved in this sector.* The World Travel and Tourism Council, 1991.

> *If we continue to treat nature as we have so far, we will undermine the foundations of our own life... we live in a postsocialist, and, I believe, a postcapitalist age. A social and ecological market economy is something quite different.* Professor Hans Kueng, Catholic Theologian, University of Tuebingen, 1991.

Over the last 20 years the concept of sustainable tourism has been developed to counter the threats which unmanaged tourism can bring. Sustainable tourism sees tourism within destination areas as a triangular relationship between host areas and their habitats and peoples, holidaymakers, and the tourism industry. In the past, the tourism industry dominated the triangle. Sustainable tourism aims to reconcile the tensions between the three partners in the triangle, and keep the equilibrium in the long term. Sustainable tourism aims to minimise environmental and cultural damage, optimise visitor satisfaction, and maximise long-term economic growth for the region. It is way of obtaining a balance between the growth potential of tourism and the conservation needs of the environment (see Krippendorf *et al.*, 1988; English Tourist Board, 1991; Hawkes & Williams, 1993; Bramwell & Lane, 1993).

Over a similar time-span, tourism in rural areas has grown, partly because of market forces, seeking different kinds of holiday, and partly as a result of government initiatives (Lane, 1991a; Weiler & Hall, 1992). This growth has been most noticeable in the countries of the developed world, where sophisticated economic diversification agencies have been hard at work promoting new uses for the countryside, influencing both potential providers of tourism facilities, and the markets for rural tourism through press and media contacts. Tourism is seen as an agent for rural economic re-generation and as a way of valorising conservation (Bramwell, 1990; Jamieson, 1990; Brown & Leblanc, 1992). The rural environment is, however, a very fragile one. It is easily either changed or damaged (or both) by rapid changes of any sort: tourism is a powerful agent for change. This is an important issue because of the role rural areas play in many nations as repositories of both natural and historical heritage. It is also important commercially. Surveys show that 'rurality' is a unique selling point for holidays in the countryside. Customers look for high quality and 'unspoiled' scenery, for peace, quiet, and, to some extent, solitude, and for the personal attention which small-scale tourism enterprises can offer to their guests (Krippendorf, 1987; P.A. Cambridge Economic Consultants, 1987). Tourism growth can be an urbanising influence, which by destroying rurality (or the illusion of rurality), can induce the onset of the destructive Resort Cycle much discussed in tourism circles (Butler, 1980).

The case for sustainable tourism in rural areas is, therefore, a very strong one. The concept of sustainability in rural tourism must be a multi-purpose one if it is to succeed. It cannot be successfully based on a narrow pro-nature conservation ethic. It should aim to:

(1) Sustain the culture and character of host communities.
(2) Sustain landscape and habitats.
(3) Sustain the rural economy.
(4) Sustain a tourism industry which will be viable in the long term — and this in turn means the promotion of successful and satisfying holiday experiences.
(5) Develop sufficient understanding, leadership and vision amongst the decision-makers in an area that they realise the dangers of too much reliance on tourism, and continue to work towards a balanced and diversified rural economy.

There are, however, many difficulties in implementing the principles of sustainability discussed above in a free market economy. The very concept of sustainability is fraught with ambiguity. Implementation is equally problematic: it involves numerous interested parties, with diverse aims and beliefs. One practical way forward lies in the creation of Sustainable Management and Development Strategies. This paper discusses the reasons for writing these strategies, their main features, and how they can best be produced. It draws on the work of the Rural Tourism Unit in the University of Bristol, over the years 1987–1993. Over that period the Unit has worked with and has examined rural tourism initiatives throughout Europe and North America, and in Australia. It has assisted the development of a number of rural tourism strategies, in Britain and in Canada. Specific reference will be made in this paper to the Sustainable Tourism Strategy drawn up for the large rural district of Berwick-upon-Tweed, situated on the Anglo-Scottish border, with a population of 28,000 people spread over 1500 sq. kms.

Why Produce Sustainable Tourism Strategies?

Almost all successful businesses, and many successful regions, develop according to carefully worked out business plans and strategies. The plans seek to reconcile competing demands, to avoid wasteful investments and duplication of effort, and to research and seek out niche markets where special success can be achieved. Sustainable Tourism Strategies should have all these attributes, but in addition, ten special features should be noted.

(1 The development of the strategy should be used to encourage an ongoing dialogue between government, tourism businesses, communities and other interests about the future of an area and tourism's role in that future.
(2) The strategy can be used to guide and encourage infrastructural investment in transport, public services, marketing, information and interpretation.
(3) Tourism businesses should gain in security and can invest for the long term because they can develop in a more stable environment.
(4) Nature Conservation, the Arts and Cultural Activities can be drawn into the strategy-making process. They will gain in stature from being recognised and valorised by being part of that process. The process should see conservation as a positive rather than a negative viewpoint. The human and political energies behind the Arts and Nature lobbies should be used to guide tourism, not simply to protect their position against tourism of any kind.
(5) The strategy can protect the special scenic, historic and cultural heritage of an area, thus preserving tourism's future seed corn. It should help to establish the financial and political interdependence between conservation and tourism.
(6) The strategy-making process can encourage new entrants into tourism. These can include farm diversification schemes, craft workshops, cycle hire, speciality food production, re-organised public transport marketed to visitors, new restaurants, etc. Often these can help retain, diversify or even increase employment opportunities. Notably, many new tourism enterprises are led by women: female activity rates have traditionally been low in rural areas (see Bouquet & Winter, 1987).
(7) Strategy-making can be used as a vehicle for new ideas and for the beginning of an ongoing educational process bringing new skills, and new flexibility into the business and political life of a region.
(8) Public dialogue and discussion can allow time for a consideration of the costs and benefits of alternative types of tourism, and alternative types of investments.
(9) Most rural tourism businesses are small and are unable to afford effective marketing or training. The strategy-making process can be used to encourage future co-operation in marketing and training between businesses and between communities, and to seek niche markets for an area where it can have an advantage over competing areas.
(10) The production of a well researched and carefully written plan, backed by the whole community, can be used as an effective lever to obtain public sector funding to begin the process of both development *and* the reconciliation of tourism and environmental issues through good management. The existence of a public sector investment programme should in turn encourage private sector investment.

But What is a Sustainable Tourism Strategy? How does a Sustainable Strategy Differ from a Conventional Tourism Plan?

Most conventional tourism plans are dominated by the growth requirements of the tourism industry. They are either marketing plans or capital investment plans. They are usually designed to increase traffic, visitor numbers, turnover, and job creation. The plans seek to exploit national and community resources and sweep away constraints. Other types of plan are often written purely in land use and physical planning terms; they may have little regard for community, culture or enterprise.

Sustainable tourism plans are based on a wider, more holistic, regional analysis. They follow the concepts first enunciated by Krippendorf, Jungk and others in the 1980s, and which since then have been refined and tested in practice in many places (see Krippendorf, 1988). Their hallmarks are:

(1) An analysis of an area's social, economic, ecological and cultural needs.
(2) An analysis of an area's tourism assets and the constraints on future tourism development.
(3) A discussion of how best tourism could be used as a tool for social, economic, ecological and cultural regeneration.
(4) A strong measure of local participation in both the plan-making process and in any ongoing decision-making.
(5) A careful consideration of the visitor carrying capacities, and the type and scale of developments suitable for each part of the region. This may include decisions to allocate some areas for intensive use, to make some areas less accessible on conservation grounds and to set overall limits to growth. Fragile ecosysytems should receive special attention; this could mean that mechanised facilities, such as ski-lifts, are excluded from key zones.
(6) A market survey to establish opportunities and a marketing strategy. This strategy should both explain and use the Sustainable Tourism Strategy as part of its campaign.
(7) A training and career enhancement programme for both businesses and employees. Part of this programme should aim to encourage new local businesses in the tourism sector; without this special encouragement and help, there will be a danger that entrepreneurs from outside the area will dominate the tourism industry before local people can adapt to the new opportunities which are available.
(8) A quality assurance programme designed to develop visitor knowledge of the area, increase insight and satisfaction, improve customer care and help ensure repeat visits.
(9) The setting up of local groups to monitor, advise, and assist future management.
(10) Stress on the need to support farming (because farmers are the guardians of the rural landscape), and on the need to encourage discussion on ways to broaden the regional economy to avoid total reliance on the visitor industry.
(11) Strategies should be long term in their approach (five to ten years appears to be appropriate), but, to be effective, must list priorities and suggest an annualised work programme for at least three years. There must be quantifiable targets, an evaluation process and regular up-dating of objectives.

How Can Sustainable Tourism Strategies be Produced?

There are four keys to success. Firstly, it is important that the person or team formulating the strategy is skilled not only in tourism development but also in economic, ecological and social analysis. While local knowledge is useful, impartiality is much more vital if trust is to be maintained amongst the many parties taking part in the strategy-making process.

Secondly, wide consultations amongst all interest groups are essential. These consultations will include trade and business, transport, farmers, administrators, and the custodians of the natural and historic assets of the area.

Thirdly — *glasnost* — openness, has a very special role to play. Tourism relies more than any other industry on local goodwill — the ability to make holidaymakers feel welcome. The local population must be happy with their visitors, and secure in the knowledge that the visitor influx will not overwhelm their lives, increase their housing costs, and impose new and unwelcome value systems on them. Openness can be achieved by public discussions, by careful use of the press, radio and television, and by the development of a two-way dialogue with the community.

Fourthly, the strategy-making process should not be a once-only affair. It has to be an evolving long-term enterprise, able to cope with change, and able to admit to its own mistakes and shortcomings. It is the beginning of a partnership between business, government and cultural and conservation interests.

It can be claimed that these keys to success are counsels of perfection. Certainly, it is probably impossible to be totally impartial, or that openness can ever be complete. But these are the ideals for which the strategy maker should strive, and ideals which should be encouraged.

The Berwick Sustainable Tourism Strategy: A Case Study

Many areas in many parts of the world have experimented with the writing and implementation of tourism strategy plans which are both environmentally and community friendly. Although the origins of sustainable tourism as a concept seem to lie in Alpine Europe in the 1970s, ideas of alternative forms of tourism in a rural milieu were tried over much of the developed world throughout the 1980s. Canada was especially active , with extensive programmes of community driven rural tourism programmes in the western parts of the country (see Murphy, 1985; Alberta Tourism, 1988, 1991; Saskatchewan Tourism, 1988). The United States of America was also involved in rural tourism development, often using University or Agricultural Extension classes as a catalyst (see Blank, 1989; Sem, 1989). Keane & Quinn's 'Rural Development and Rural Tourism' (1990) provides a detailed review of Irish developments, as well as an overview of work in continental Europe. A full bibliography of rural tourism in the developed world has been prepared for the OECD (Lane, 1993).

The case study given here typifies and exemplifies many of the ideas and problems in the development of rural tourism strategies commonly found within the developed world. The Berwick district is one of the largest and most sparsely populated administrative areas of England. Over 500 kms north of London, it contains the mountain landscape of the Cheviot range, the rolling Kyloe Hills, the wide valley of the River Tweed, and 50 kms of sandy coast line (see Figure 1).

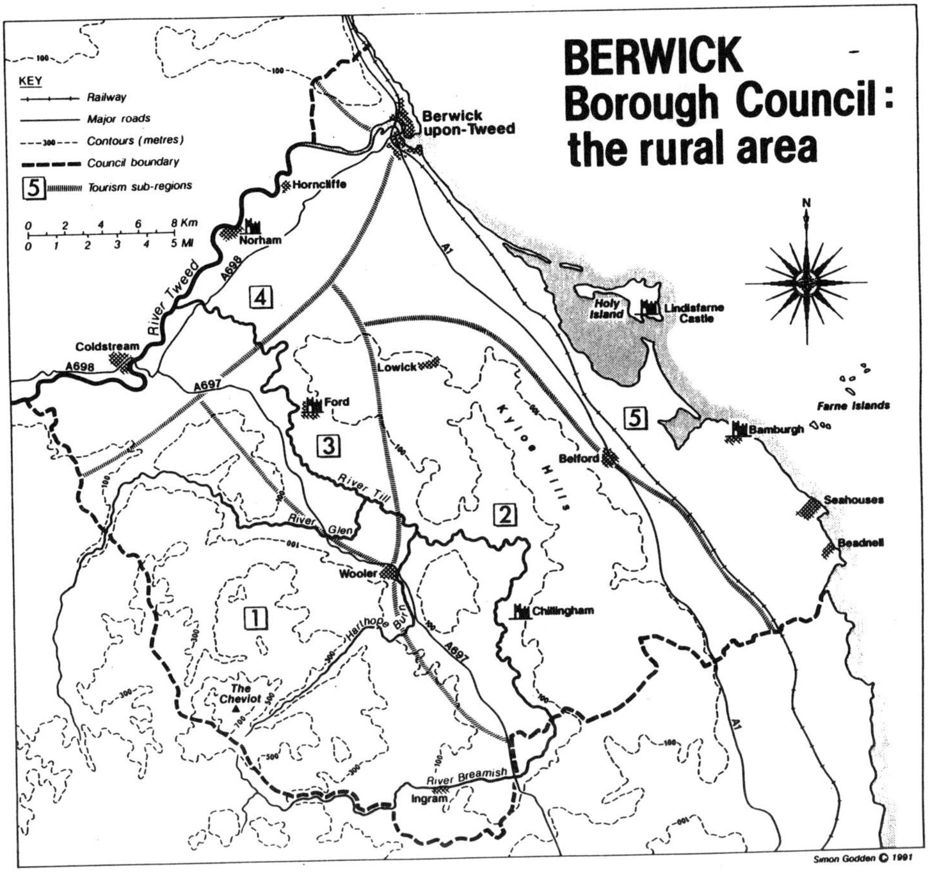

Figure 1 Berwick district (*Key*: Tourism sub-regions: 1. Cheviot; 2. Kyloe Hills; 3. Glendale; 4. Tweed Valley; 5. The Coast)

There is an interesting historic heritage, an important element of which reflects medieval conflicts between England and Scotland. The economy is dominated by farming and rural services. Between 1851 and 1981, Berwick's rural area lost 53% of its population; over the same period, the total population of England and Wales rose by 188%. Job opportunities in the area remain poor, and out-migration continues. As a result, the percentage of people of pensionable age, 23%, is above the national average of 17%; some parishes have over 35% of their populations over 65 years old.

Tourism has been slow to develop, except to a limited extent along the coastline. The local council has now appointed a Tourism Officer and begun to market the area. In 1989 the University of Bristol was asked to advise the council about setting up a heritage centre in its second largest town, Wooler. After discussions, it was agreed that this idea would be best considered by developing a Sustainable Tourism Strategy for the whole Berwick Borough Council area.

The work began in early 1990, and was completed in July 1991. A series of desk studies assessed the economic, social and ecological issues in the area in some depth. This was followed by a visual inspection of the entire region, and discussions with all interested parties — over 50 in all. In May and June, 1990, a series of public meetings was held in each of the nine main communities. They were widely advertised, and timed to secure maximum attendance. The meetings were open — but carefully structured to:

(1) Explain the strategy-making process.
(2) Ask participants how they saw the future of their communities: they were asked to discuss both the best and worse scenarios they could envisage in ten years time.
(3) Ask the participants how they viewed the possibility of tourism development.
(4) Outline to the communities the different types of rural tourism development which were possible in their communities.

These meetings were often long, sometimes noisy, but always useful. They were open but carefully structured to help communities begin to consider in detail where their futures lay. They were the beginning of a dialogue. They were also the beginning of a learning curve for many of the participants — about tourism, and about decision-making in general. Crucially, they revealed which communities had the human energy and drive to cope with change and with visitor management. Each meeting was reported at length in the press and on local radio.

A draft plan was produced for comment over the winter of 1990/91, and a further series of structured public meetings were held in the main communities in Spring 1991. Again there was press, radio and television coverage. Amendments and improvements were then made to the strategy which was approved by the Berwick Borough Council in September 1991 (for the full strategy document, see Lane, 1991b).

What did the strategy say?

In summary, it recommended that:

(1) The area had neither the resources, nor the wish, to be developed as a mass tourism area: it should therefore capitalise on this situation and develop small-scale markets seeking heritage and closeness to nature.
(2) Conservation and tourism management should be the first priority, followed by job retention and creation and the encouragement of business, especially locally based businesses.
(3) One area, Glendale (see Figure 1), should be zoned for relatively intensive development. It was ecologically strong, and its service centre, Wooler, the second largest in the area, needed economic support if it was to retain a full range of shopping and other facilities.
(4) The Cheviot mountain wilderness should be a zone where access was made more difficult because of its fragile ecosystems. Alternative walking routes should be developed along the edge of the area to satisfy visitor demand while safeguarding the central dome of the Cheviot. An attractive interpretation centre in Wooler would explain this policy and help redistribute visitors to Glendale, the Tweed Valley and the Kyloe Hills. Measures should, however, be taken to assist hill sheep farms to undertake limited and care-

fully controlled diversification into tourism; without this help there is a danger that the Cheviots will be sold for forestry purposes. This would change its character completely, to the detriment of existing ecosystems, and to the marketing image of the whole region.

(5) The Tweed Valley should see bicycle tourism and river walks developed for a family market and for older people. These activities should be focused on the village of Norham, in order to support its shopping and other services. Special provision should be made here for the better use of public transport; the use of cars should be discouraged on some roads.

(6) The Kyloe Hills should be developed for walking and farm tourism, because they are ecologically strong, and their sheep/beef cattle farming communities need economic support and diversification. The hills need a comprehensive footpath and parking policy, and a marketing initiative.

(7) The coastal area should be a zone of restraint on new development. But efforts should be made to build a series of small landscaped car parks to encourage birdwatching and short circular walks. Measures should also be developed to guide visitors towards the small fishing harbour at Seahouses. The fishing industry is suffering economic decline in most parts of Britain. Seahouses should, therefore, be the subject of a co-ordinated regeneration plan to improve its housing stock, its amenities, and its economic base. An improved contribution from the tourism sector will help, but other initiatives are also needed.

(8) Local development groups should be formed to ensure grassroots involvement. The first of these could be in the small towns of Belford and Wooler, and in the village of Norham.

(9) A comprehensive footpath, signpost, information and interpretation policy should be implemented, designed to help visitors get to know the area, understand its wildlife and its historical heritage, stay longer and spend more money.

(10) A new more closely focused marketing plan should replace the current broad approach. Marketing would be targeted towards special interest groups and specific geographical areas.

(11) A series of training and education courses should be offered to encourage new entrants to tourism and to help existing businesses.

(12) A major investment in infrastructure and staff should be made in order to fund both development and management. The investment should be phased over an initial three years, followed by a seven-year follow-up programme. Part of the investment should be made in an ongoing process of strategy monitoring and evaluation, adapting the strategy to changing conditions.

Funding totalling £450,000 over the first three years was sought by Berwick Borough Council on behalf of the communities and businesses of the area during the latter part of 1991 and early 1992. A joint public sector/private sector partnership was sought: the public sector was to contribute the majority of the funding in the early stages of infrastructure development, until there was sufficient progress for private sector funding to be brought into play.

Funding negotiations proved to be slow-moving. It was necessary to assemble a partnership of public sector bodies: this process was difficult because of the differing aims and requirements of the bodies involved. The private sector was unhappy about investment plans at the onset of the major recession which overtook Britain in this period. Berwick Borough Council is one of the smallest English local authorities in terms of population size and, therefore, has limited staff resources. It found the negotiations extremely demanding in terms of special skills and staff time; problems were exacerbated when the Chief Executive, who had been leading the funding bid, left to take up a more senior post elsewhere.

It took until April 1994 for the major funding breakthrough to take place, and the resources then secured were less than originally requested The lesson to be learnt from this is that strategy funding can be a task as difficult and time consuming as the writing and implementation of the strategy itself.

Conclusion

The processes outlined in this paper are long and complex. They require detailed and painstaking work. The ideas behind Sustainable Tourism Strategies are frequently denigrated by people in a hurry. They are sometimes attacked by developers seeking rapid, short-term returns on their investments. Other regions may follow conventional short-term plans, and gain more jobs and more visitors and more investment. But pause. If the people of an area love their landscape, their way of life, their language, their dialect, their cuisine and their heritage — they will pause. They will reflect that the creation of the great palimpsest that is the countryside has taken many years of human effort. They will remember the toil of those first settlers who cleared the woodlands, those who built farms, villages, roadways, labouring to add churches, schools and gardens, founding families and traditions. What are perhaps 18 months of thought and discussion, followed by probably 10 years of carefully managed change, in that long history? Organic change, reconciling tensions between conflicting motives, needs time. Managed change need not be obtrusive: strategy makers would do well to remember the words of Sir Joshua Reynolds, the first President of the Royal Academy, 'Art is to conceal art' (Reynolds, 1797). Perhaps that could be a watchword for practitioners of sustainable tourism generally…unless they work in the marketing department.

References

Alberta Tourism (1988) *Community Tourism Action Plan Manual.* Edmonton: Alberta Tourism.

— (1991) *Market Planning Skills Program: Community Participants' Guide.* Edmonton: Alberta Tourism.

Blank, U. (1985) *The Community Tourism Industry Imperative: The Necessity, the Opportunities, its Potential.* Pennsylvania: Venture Publishing.

Bouquet, M. and Winter, M.(1987) *Who From Their Labours Rest? Conflict and Practice in Rural Tourism.* London: Gower.

Bramwell, B. (ed.)(1990) *Shades of Green: Working Towards Green Tourism in the Countryside.* London: English Tourist Board.

Bramwell, B. and Lane, B.(1993) Sustainable tourism: An evolving global approach. *Journal of Sustainable Tourism* 1 (1), 1–5.

Brown, V. and Leblanc, F.(1992) *The Ontario Heritage Regions Projects.* Ottawa: Heritage Canada.

Butler, R.W. (1980) The concept of a tourist area cycle of evolution: Implications for the management of resources. *Canadian Geographer* 24, 5–12.

P.A. Cambridge Economic Consultants (1987) *A Study of Rural Tourism.* London: Rural Development Commission/ English Tourist Board.

English Tourist Board (1991) *Tourism and the Environment: Maintaining the Balance* (plus four associated volumes: the reports of the Countryside Working Group, the Heritage Sites Working Group, the Historic Towns Working Group, and Visitor Management Case Studies). London: English Tourist Board.

Hawkes, S. and Williams, P. (1993) *The Greening of Tourism: From Principles to Practice.* Burnaby, British Columbia: Simon Fraser University Press.

Jamieson, W. (ed) (1990) *Maintaining and Enhancing the Sense of Place for Small Communities and Regions.* University of Calgary.

Keane, M.J. and Quinn, J. (1990) *Rural Development and Rural Tourism.* University College Galway.

Krippendorf, J. (1987) *The Holiday Makers: Understanding the Impact of Leisure and Travel.* Oxford: Heinemann.

Krippendorf, J., Zimmer,P., and Glauber, H.(1988) *Fuer einen andern Tourismus.* Frankfurt: Fischer Taschenbuch Verlag.

Lane, B. (1991a) Will rural tourism succeed? In S. Hardy, T. Hart and T. Shaw (eds) *The Role of Tourism in the Urban and Regional Economy* (pp. 34–39). London: Regional Studies Association.

— (1991b) *Berwick-upon-Tweed: A Strategy for Sustainable Rural Tourism.* University of Bristol.

— (1993) *Rural Tourism: A Bibliography.* Paris: OECD.

Murphy, P.E. (1985) *Tourism: A Community Approach.* London: Methuen.

Reynolds, Sir Joshua (1797) *Discourses.* London. (Modern edition ed. R.R. Wark (1975), New Haven, Yale University Press).

Saskatchewan Economic Development and Tourism (1988) *Community Tourism Planning Guide.* Regina: Saskatchewan Tourism.

Sem, J. (ed.) (1989) *Using Tourism and Travel as a Community and Rural Revitalisation Strategy (Proceedings of the National Extension Workshop).* Minneapolis: University of Minnesota.

Weiler, B. and Hall, C.M. (1992) *Special Interest Tourism.* London: Belhaven/Halsted Press.

New Supports for Heritage Tourism in Rural Ireland

Adrian Phillips and Mary Tubridy
Natural Resources Development Centre, 188 Pearse Street, Trinity College Dublin, Ireland

National Development Plans for the 1990s in Ireland identify tourism as a sector with significant potential for job creation, particularly in rural areas. A new tourism policy has been developed promoting the presentation of the natural and human heritage of the country and development of outdoor activities. This will be supported during the 1990s by an investment of about Ir£2 billion from EC, national and private enterprise sources. The success of this investment in producing sustainable rural tourism will depend upon careful planning which integrates heritage-based tourism within broader planning of integrated rural development and of conservation of landscape and human heritage. Appropriate marketing methods must also be developed.

This paper presents examples of some new concepts and practices in the planning, interpreting and marketing of heritage-based rural tourism in Ireland. These include uses of Geographic Information Systems, methods of involving rural communities in interpreting their heritage, attempts to promote community based rural development which is balanced with conservation, and the use of digital techniques, including networked interactive computer multimedia systems for marketing heritage-based tourism in Ireland.

Introduction

The most serious economic problems facing the Republic of Ireland at present are unemployment, with a national average of more than 20% of the work force, and the threat to rural communities from progressive reduction in EEC subsidies for agriculture. National development plans for the 1990s identified tourism as an area with potential for creating significant numbers of new jobs in both urban and rural communities. With an investment of Ir£750 million from national, EC and private enterprise funds, 25,000 new jobs were created in the tourism industry during the period 1988–93. In the National Development plan for 1994–9, a total of Ir£1.6 billion, including Ir £500 million from EC, is targeted to generate a further 23,000 new jobs in tourism. In response to market surveys which showed the popularity of heritage attractions (e.g. in 1991, 62% of the 4.6 million visits to Irish heritage attractions were made by overseas visitors), particular emphasis has been placed by the Tourist Board (Bord Failte) on the development of heritage and activity based tourism. In 1990 a consultancy report commissioned by Bord Failte concluded that the quality of heritage attractions in Ireland was low by international standards, that there was a great tendency to duplicate the same themes, and little success was achieved in giving visitors a comprehensive understanding of Ireland's heritage. In attempting to rectify these weaknesses, Bord Failte has produced a strategy to interpret for tourism Ireland's natural heritage, history and culture. The strategy aims to assist in developing a coherent national scheme of interpretation in which duplication is avoided, but networking of areas to present an overall theme is encouraged. This is summarised in 'Heritage and Tourism', the Proceedings of the second conference on the Development of Heritage Attractions in Ireland organised by Bord Failte at the Royal Hospital Kilmainham on 28–29th January 1992. The strategy involves clustering storylines under the following five themes:

Live Landscapes
Making a Living
Saints and Religion
Building a Nation
The Spirit of Ireland

Each theme is intended to provide a focus around which numerous storylines can be developed. The following framework of themes and major storylines is proposed:

THEME	STORYLINES
(1) Live Landscapes	(a) Land and Sea
	(b) Mountain and Moorland
	(c) Cliffs and Caves
	(d) Bogs and Wetland
	(e) Rivers and Lakelands
	(f) Wildlife and Nature
(2) Making a Living	(a) Emigration and Famine
	(b) Working with the Sea
	(c) Industry, Transport and Power
	(d) Inventions, Transport and Power
	(e) Lifestyles through the Ages
	(f) Farming Heritage
(3) Saints and Religion	(a) Pagan Ireland
	(b) Early Christianity
	(c) Saints and Scholars
	(d) Missionary Spirit
	(e) Religious Upheaval
	(f) The Living Faith
(4) Building a Nation	(a) Celts and High Kings
	(b) Invasion and Conquest
	(c) War and Rebellion
	(d) The Anglo-Irish
	(e) People and Places
	(f) Ireland's Story
(5) The Spirit of Ireland	(a) Literary Ireland
	(b) Language, Folklore and Legend
	(c) Ireland Entertains
	(d) Art in Ireland
	(e) Irish Games and Sport
	(f) Famous Personalities

The storylines are not exhaustive and new ones are possible under all themes.

The successful implementation of these ideas and the wise spending of the large level of funding currently available pose a major challenge over the next five years. Very careful and coordinated planning from regional to county and down to individual project level is needed. The best information on Ireland's heritage must be drawn out from academic and community sources and presented in imaginative and entertaining forms. This must be balanced with conservation of areas through such devices as the development of National Parks, of Nature Reserves and of Environmentally Sensitive Areas, interlinked with evolving plans for integrated rural development. This paper presents approaches which are being used at Trinity College Dublin as aids to the development of sustainable niche opportunities in tourism for rural areas. These involve:

(1) Promoting community-based tourism.
(2) Helping to focus tourism in rural areas on heritage which is unique to each area.
(3) Promoting the sustainability of rural tourism by assisting the development of plans that link tourism into broader plans for integrated rural development.

(For a broad discussion of the issues involved in Ireland beyond the work of this paper, see Curtin & Wilson (1988), Feehan (1992) and Keane & Quinn (1990).) The overall aim has been to promote the development of a stable balance between tourism, primary production (e.g. farming, forestry, fisheries, mining and quarrying), manufacturing and service industries, and conservation of the natural and human heritage. Each area will have a different optimum balance between these activities, which will change with time. Our contribution from the Natural Resources Development Centre at Trinity College has been to assist in the following activities:

(1) Production of *resources databases* for an area, usually in digital form using a Geographic Information System (GIS), of basic characteristics of the natural environment (geology, economic deposits, soils, topography, river systems, fauna and flora), and of the man-made heritage (archaeology, architecture, history, literature, music, sculpture, socio-economic census data).
(2) Production of *tourism databases* covering information on existing facilities and attractions for visitors, and the numbers, origin, interest, timing of visits and level of satisfaction of visitors at existing tourism attractions in an area.
(3) Use of the databases (1 & 2) for the production of an *interpretative plan* defining what, where and how the natural and human heritage and present day activities in an area should be presented to visitors.
(4) Inclusion of tourism in *integrated development plans* for an area through a consultative process involving the community, statutory organisations, specialist expertise and user groups. The aim of such plans is to maximise conservation of the population through job creation (including tourism) and improvement to their quality of life in ways which are compatible with conservation of the natural and human heritage of the area.
(5) Provision of commercial services in the research for, design of, and implementation of interpretative displays, including the production of films and

audio-visual systems. These services have been provided through the establishment of a campus company at Trinity College Dublin.

(6) Production of marketing materials and systems. This has also involved a commercial campus company service in digital production of high quality maps and brochures, and the production of interactive computer multimedia information systems for marketing tourism services and heritage attractions.

We use examples of our work in Counties Wicklow and Wexford (Figure 1a) to illustrate the provision of these planning, production and marketing services as aids to the development of opportunities in sustainable niche tourism. By *niche tourism* we mean a tourism business which is stably rooted in the regional economy, contributing significantly to the well being of the community and dependent primarily upon all that is unique in the natural and human heritage of the area.

In County Wicklow, the presence of a population of over 1 million people in the adjacent area of Dublin city presents an opportunity through having easy access to a large market for products and services and to international transport services. However there are also threats to rural development and niche tourism

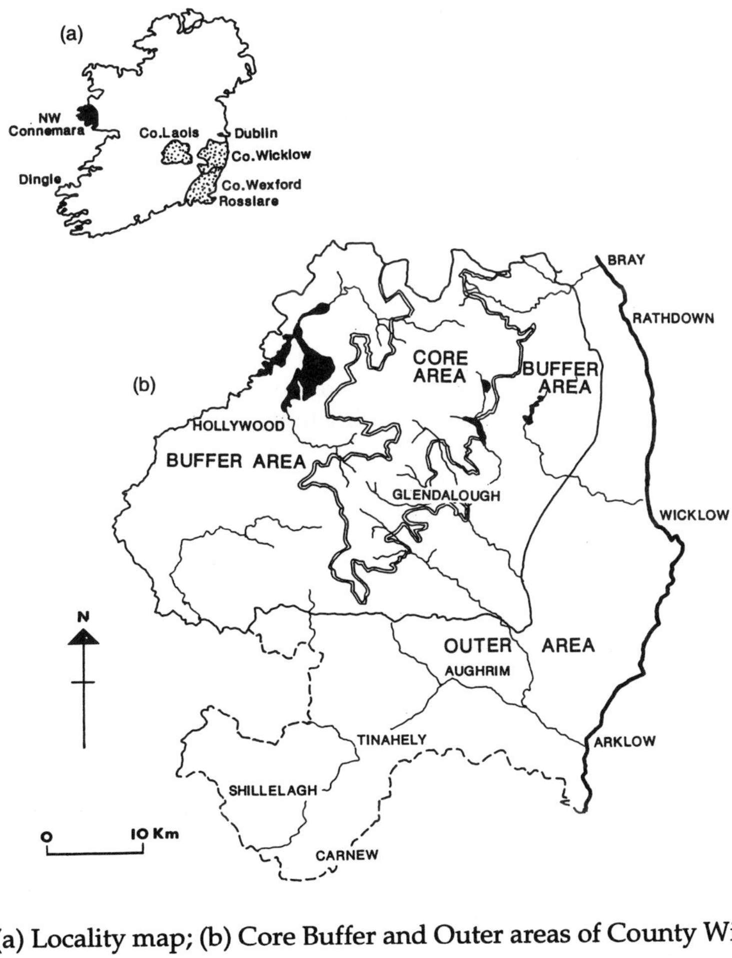

Figure 1 (a) Locality map; (b) Core Buffer and Outer areas of County Wicklow

arising from the county at present being used predominantly as an unpaid for park for day trips from the city, and from the inflation of land and property prices in the county as it becomes a commuter suburb for more prosperous Dubliners. There is a particular need for an increasingly disadvantaged rural community to become more fully involved with tourism if they are to survive. There is also a need for the planning and development of a newly established Wicklow Mountains National Park to be fully integrated and balanced with plans for rural development in the county.

The ferryport at Rosslare in County Wexford is one of the main entry points for visitors from Britain and mainland Europe. However most of these visitors pass through the county without staying a night. The challenge is to persuade these migrant visitors to stay in the county to enjoy the wealth of scenic attractions and of historical heritage.

Production of a Resources Database: County Wicklow

In May 1990, Government plans were announced for a Wicklow Mountains National Park which is planned to include 30,000 hectares in the core area of the mountains, lying generally above the 400 metre contour. This led to a growing level of concern about potential negative consequences for sheep farmers and residents adjacent to the proposed park. In September 1990, as a response to this reaction, Wicklow County Council commissioned the Natural Resources Development Centre at Trinity College to develop a management strategy for sustainable development and conservation in the Wicklow Uplands area (Tubridy *et al.*, 1992). The aims of the Wicklow Uplands Study were:

(1) To produce an integrated development strategy for the Wicklow Uplands region which balances the need for economic development with the need to conserve the outstanding qualities of the landscape and heritage resources of the area.
(2) To identify actions which are needed to enable the uplands of Wicklow to become a renewable resource stimulating social and economic development in the region and not a rapidly diminishing and non-renewable resource.
(3) To ensure that the proposals which emerge arouse widespread support and make practical sense by involving widespread consultation and discussion. This entailed holding a series of public meetings and conferences and also discussions with individuals and organisations involved with the area.

The first phase of this 'Wicklow Uplands Study' involved the production of a comprehensive database of resources, using an ARCINFO GIS system to store relevant data covering the natural and man made heritage, and the current activities and socio-economic census data for the area. By using the ARCINFO software and CENMAP (a new software system developed by Gamma Ltd for graphical display of census data), it is then possible to create composite maps showing for example areas suitable for farm-based heritage tourism. Such a map involves use of socio-economic data, of data on present farming activities, archaeological sites, walking routes, scenic areas, distance from tourist facilities and attractions etc. Illustrating the widespread utility of such a data base in rural development, the Department of Microelectronic Engineering at Trinity College is currently using it in conjunction with ray tracing software and ground signal measuring equipment to optimise designs for the mobile cellular radio communication systems in County Wicklow.

Production of a Tourism Database: County Wicklow

In response to the need to develop a stronger tourism industry in the county, rather than a day trip one, and to promote employment, the County Wicklow Tourism Authority commissioned a consultancy study entitled 'Tourism in Co. Wicklow — Maximising its Potential' (Convery *et al.*, 1989). This includes a tourism database in the form of a valuable summary of the strengths and weaknesses of the existing resources such as: accommodation, inland waters, coastal waters, countryside, leisure centres, sporting and cultural activities, restaurants, entertainments, agritourism and infrastructural features. A survey of demand for tourism facilities in County Wicklow was also carried out in Dublin.

A major aspect of our subsequent work in preparing a management strategy for the Wicklow Uplands and an interpretative plan for the county has focused on identifying ways of creating a sense of place by which local communities and their visitors can gain more benefits from tourism in their areas. This has involved adding to the tourism data base (section 2) detailed information on the natural environment, the man made heritage, the cultural heritage, and on life in the area today. A similar database has recently been produced (Tubridy, 1993) for the NW Connemara area (Figure 1a) where a particularly disadvantaged rural community also need to share much more fully than at present in the tourism industry of that area.

Development of an Interpretative Plan

The purpose of an interpretative plan is to ensure that the right level of resources is allocated to interpretation projects which meet the needs of visitors, reflect the interest of local communities, are appropriate in nature and scale to their location, avoid duplication of information and services, and are well planned and implemented.

Development of an interpretative plan for County Wicklow

Interpretation systems are needed at present in County Wicklow to address the urgent problems of managing pressures associated with about two million day visitors per year in the Uplands and of promoting employment through rural tourism.

The development by the state of a National Park and of visitor centre facilities offers great potential benefits for the area, provided that their planning involves a strong focus on consensus building with the rural community, and that these plans are integrated within wider development and conservation plans for the county.

As a result of its landscape of outstanding quality adjacent to the capital city, and because of the highly complex ownership situation in the Uplands, County Wicklow poses unique problems in Ireland for the development of a National Park. This special situation means that particular emphasis must be given to achieving a sustainable balance between conservation and rural development. In considering how to achieve this, Wicklow County Council commissioned a study to develop an interpretative plan for the whole county in which the plans for the National Park and its interpretation were integrated in a larger perspective (Phillips *et al.*, 1992). The aims of the plan were:

(1) To increase visitor satisfaction and maximise social and economic returns to local people, consistent with conservation of the natural and social environment.
(2) To maximise the involvement of local communities in planning and implementing interpretative facilities.
(3) To support recreation, tourism and community development policies of public, private and community organisations.
(4) To give guidance to public, private and community organisations on the range and type of interpretative facilities and services which would be appropriate.
(5) To provide a framework for funding agencies for responding to initiatives and setting priorities.
(6) To provide guidance on the process of developing and implementing schemes.
(7) To prepare a strategy which will be the basis for coordinated application for the EC structural funding under the Operational Programme for Tourism, and a framework for responding to specific proposals from individuals and organisations in the county.

Methods

The production of an interpretative plan for County Wicklow was carried out with support from staff of the Centre for Environmental Interpretation at Manchester Polytechnic. The project involved a seven man-month study in which there was exhaustive consultation with individuals, community organisations, and officials in the local authority and in relevant state organisations. It was the first attempt to use, on a county-wide level, the new Bord Failte strategy of themes and storylines in heritage interpretation. This involved drawing together by research and widespread consultation an integrated plan which defined all the themes and story lines which were unique to each part of the county, avoiding duplication by identifying interpretation networks, and outlining where and how interpretation of themes and storylines should be developed.

The county was divided into three regions, the *Core Area* consisting of the proposed 30,000 hectare National Park. The *Buffer Area* surrounds this and generally lies above the 300 metre contour. The remainder of the county, including the coastal and south Wicklow areas, forms the *Outer Area* (Figure 1b).

Themes and storylines for the Core Area: Wicklow Mountains National Park

Live Landscapes

A major theme of the Core Area is the landscape and its relationship to natural and human life. This fits into the Bord Failte theme of *Live Landscapes*. The unique features of the landscape of the Core Area are the mixture of rounded topped granite mountains with their peat and moorland vegetation cover, local corrie lakes, and the wildlife of birds and mammals (e.g. deer). In the west there are broad valleys with woodlands, green fields and small farms. In the east the rivers are more deeply incised with lakes and waterfalls. These characteristics can usefully be divided into storylines of *Mountains and Moorlands; Valleys and Lakes; Plants, Trees and Wildlife*. These three storylines fit with minor modification into the storyline structure suggested by Bord Failte for the theme of *Live Landscapes*.

The human themes of the Core Area fit readily into the Bord Failte themes of *Making a Living; Saints and Religion;* and *Building a Nation*.

Making a Living

Under the theme *Making a Living*, the following storylines are important:

(1) The role of Stone Age people in creating passage graves on the tops of some of the mountains and their possible role in accelerating the growth of blanket peat by burning and overgrazing.
(2) The history of mining lead, silver and copper.
(3) Sheep farming in the Uplands.
(4) Production of hydroelectric power in the Uplands.

Saints and Religion

Within the theme of *Saints and Religion*, the *Early Christian* and *Saints and Scholars* storylines can be combined to cover the early Christian monastic sites (Glendalough, Hollywood) and their connection by an ancient pilgrim's road.

Building a Nation

Under the theme of *Building a Nation*, there are two important storylines:

(1) Under *War and Rebellion* comes the 1798 Rebellion, including the use of the remote valleys as assembly points and refuges by the rebels, and the response by the government of the time which included the building of a military road linking a series of barracks in the Wicklow Uplands.
(2) The management of the Core Area in the period of the big estates during the 18th and 19th century fits into the *Anglo-Irish* storyline.

Interpretation strategy for themes and storylines in the Core Area: Proposed Wicklow Mountains National Park

Of primary importance in designating an area such as the Core Area of the Wicklow Uplands as a National Park is that there should be provision of public access to it, and that there should be facilities to allow full appreciation of it. In response to these requirements, the Office of Public Works decided to invest about £4 million on an interpretative centre in a particularly scenic place within the Core Area. This proposal was met by widespread controversy. An alternative strategy is proposed in the Interpretative Plan. It was recommended that in a *Phase 1*, available funding should be diverted to installing a vital infrastructure of facilities and services in the Uplands. Firstly the infrastructure should consist of a very high quality 'interpretative system' for the Wicklow Uplands. This should consist of a comprehensive data base on the natural and human heritage of the area to be drawn up by National Park staff working in close cooperation with communities, academics and a series of community training courses on heritage in rural tourism (see below). This material could then be used to produce a range of interpretative materials such as thematic maps, brochures, guide books, displays, exhibition material, audio-visual presentations and interactive multimedia displays. Secondly, the infrastructure should consist of a network of 'focal points' consisting of car parks with a small stone building to provide shelter for an attendant and to house appropriate on site interpretation, such as a display map of the area, guide booklets and walkers maps, managed walking routes, warden services, manned car parks, and local walker bus services. Their function is to provide adequate access and appreciation so that the area can attain the status of an internationally acceptable National Park. They are also required to address the urgent need for interpretation systems and access facilities to reduce existing serious recreational pressures and growing problems of theft and vandalism, particularly in the northern half of the Uplands close to Dublin. This infrastructure could be used to provide, for the first time, market research information on who uses the Uplands for what purposes and when. Such information is an essential foundation upon which plans for large investments in providing visitor centre facilities, which are sustainable, can be drawn up and implemented.

As a result of experience gained in *Phase 1* it will be possible to define the target markets, visitor numbers, timing of visits, modes of transport and on-site interpretation services required at major visitor centres. This should then be the basis for a further investment in providing centres.

In *Phase 2* a sum of at least Ir£5 million will be needed to develop two visitor centres adjacent to existing settlements in the Buffer Area on the east and western sides of the Core Area. Operating costs will be lower if the centres are close to existing settlements where they can have an all-year usage involving communities as well as visitors. It was also recommended that before proceeding with the construction of Visitor Centres, there should be a widespread process of consultation using the management planning committee for the Uplands (see below) recommended in the Wicklow Uplands Study (Tubridy *et al.*, 1992). Full Environmental Impact Assessment should be carried out and the projects must be compatible with and supportive of the County Development Plan.

Interpretative Plan for the Core Area

Arising from the interpretative strategy, detailed proposals for interpretation were then presented for each site in a network of 11 focal points in the Core Area.

Interpretative Strategy for the Buffer Area

In the *Buffer Area* of the Uplands, the most urgent priority is to implement, after achieving a widespread consensus, a recreation management plan which will alleviate the many serious problems caused by the existing day visitors. Larger scale interpretative facilities, including Visitor Centres for the National Park should be sited in the Buffer Area where there is the necessary infrastructure including local communities, and where it is possible for additional services for visitors to be provided by private or community enterprise. A number of themed focal points should be developed in the Buffer Area, in addition to those already there. These should be designed for 50–100,000 visitors a year.

Each major focal point should be supported by a network of trails, guided walks and interpretative panels in the immediately surrounding area.

There should also be a network of smaller sites illustrating the main themes of the area, including natural sites, historic buildings, archaeological and industrial archaeological sites, view points, and places where local life and culture can be presented and explained. These smaller sites should be interpreted using walking trails, interpretative panels, pictorial maps, guided walks and tours, local information points, family activity packs, events and demonstrations.

Interpretative Plan for the Buffer Area

Using the strategy suggested, detailed proposals were made for how and where themes and storylines concerned primarily with the interactions through time of people on the landscape should be presented at 28 sites in the Buffer area. Emphasis was also placed upon the development of walking routes along rivers which rise in the Core Area and flow through scenic valleys in the Buffer Area.

Interpretative Strategy in the Outer Area

In the coastal belt of the *Outer Area,* four very appropriate interpretative centres are at various stages of development from planning to completion (Wicklow Gaol, the Arklow Maritime Centre, Rathdown Medieval Village Centre and Bray Heritage Centre). It is suggested that a coastal walking route linking them could be developed.

Inland the area divides into smaller 'catchments' for tourism focusing around five small towns and villages, each having its own distinct character. The recommended strategy for this inland part of the Outer Area consists of:

(1) A range of small scale interpretation facilities and services based in and around small towns providing accommodation, designed to encourage visitors to stay and explore the natural and historical heritage as 'paying guests of the local community'.
(2) Themed interpretation provided by local groups and other relatively low-cost self-guided trails, local leaflets and parish maps, specially designed walks, programmes of guided walks, farm open days, woodland walks, evening talks and entertainments.
(3) A number of local information centres with multimedia displays and access to local tourist information systems including local history, landscape, genealogy, events, products and services.
(4) The development and promotion to visitors of a programme of local events, sporting activities and festivals.
(5) The development of a network of extended walking trails, cycling routes and Mass Paths with associated interpretation linked to accommodation and baggage carrying services for package holidays.

Interpretative Plan for the Outer Area

Arising from the interpretative strategy suggested, detailed proposals were presented for how and what interpretation should be presented at 33 sites in the Outer Area, which would create a sense of place in each of five centres of accommodation.

Conclusions for interpretation in County Wicklow

Tying all these initiatives together into a viable and coherent network which avoids duplication requires:

(1) Strong thematic presentation of the main stories of Wicklow's natural landscape and social and cultural history.
(2) Development of a county-wide heritage database.
(3) Creation of interpretative materials such as maps, leaflets, brochures, guide books, videos, AV, videotext and multimedia presentations.
(4) Development of a network of walking routes, which should include those which will lead people outwards from the Core Area along rivers which rise there.
(4) Development of a network of cycling routes.
(5) An events programme such as festivals, sports, artistic performances and exhibitions.
(6) A short stay visitors pack for each area which contains information, leaflets, local events programmes on what there is to do and see.
(7) Provision of county-wide information on attractions and activities at each focal point.

How to make it work

The implementation of these plans will be greatly facilitated by:

(1) The provision of two full time training courses on Heritage in Rural Tourism as outlined below.
(2) A series of short courses and experience exchange visits for decision makers, senior administrators, wardens, volunteers and part-time wardens and small businesses, accommodation owners and attraction managers.
(3) The County Development Team in cooperation with Wicklow Rural Enterprises Ltd (supported by EC LEADER Programme — see below) drawing up a panel of specialists and providing financial support towards the cost to communities or individuals of employing them at an early stage for technical advice in : (i) Interpretative planning, (ii) Feasibility studies, (iii) Design advice and costings, (iv) Visitor management and customer care, (v) Marketing and promotion
(4) Availability of public sector support for capital works, feasibility studies and implementation plans for large scale projects (EC Structural Funds, EC LEADER programme).
(5) Availability of small grants programmes for local/small business initiatives and conservation projects (e.g. LEADER)
(6) Availability of additional funding for co-ordinated marketing support (e.g. LEADER, Wicklow Tourism Authority)
(7) Creation of a forum of public, private and voluntary organisations involved in promoting interpretation in the county to: (i) Coordinate initiatives, (ii) Appraise progress, (iii) Respond to new developments and changes in funding opportunities, tourism patterns etc.

Training in Heritage Interpretation and Marketing Systems

If rural tourism is to be sustainable, it is essential that it be led by individuals and groups within the rural community who have an extensive knowledge of the relevant heritage resources of their region. To help this to develop, a series of full time six-month training courses on 'Heritage in Rural Tourism' have been provided by the Natural Resources Development Centre. These have been funded as external training courses by the National Training Authority (FAS). They have been held so far in County Laois, Dingle and SW County Wexford. The trainees are selected on the basis that they are committed to developing a business from heritage-based tourism, either as providers of accommodation, as guides or as staff of heritage centres. The syllabus involves:

(1) The first two months is spent covering a wide range of different areas of heritage, e.g. Geology, the Quaternary Ice Age, the present landscape, Fauna and Flora, Archaeology, History, Folklore, Industrial History, activities etc. The range of topics is selected to cover the themes and storylines already identified from interpretative plans for each trainees area.

(2) During the second two months, the trainees carry out, as a group, a regional survey of heritage resources in cooperation with a local community. Particular emphasis will be placed on training in how to record this data through the production of maps and drawings, by word processing, and by recording through photography, tape and video.

(3) During the third two months each trainee carries out an individual survey in their home area. This will lead to the presentation of a detailed report on its heritage, including a map showing the position of all sites, a data base in the form of text, drawings, colour slides and where appropriate tape and video recordings. The final report also includes a business plan for the use of this information in promoting a rural tourism business (accommodation service, guide service, heritage centre etc.).

An introduction is also given to good practice in graphic design and training is also provided for certification in computer keyboard skills and in elementary conversation level in at least one foreign language. These courses can play an important part in data collection for displays and for marketing materials such as maps, brochures, guide books, videos and interactive multimedia computer systems.

Inclusion of Rural Tourism in an Integrated Development Strategy: The Wicklow Uplands Study

The aim of the Wicklow Uplands Study (Tubridy *et al.*, 1992 and above) was to recommend a strategy for sustainable development and conservation in the Uplands, using a process of widespread consultation. The strategy was based upon a comprehensive account of the natural and cultural heritage, of the socio-economic conditions in the area and on analysis of the impacts of National Park policies, of forestry and of recreation. The research was used as a way of empowering communities in the Uplands to take part in the debate regarding the future development of their environment and to play a part in the development of future integrated management strategies.

Methodology of the Wicklow Uplands Study

The approach of the research team for this study, which included environmental scientists, economists, social scientists and geographers, was to emphasise the participation of the local people, to understand, reflect and record their opinions and to bring about a process of greater understanding and awareness of their situation as a basis for informed decision making. It also led to the preparation of a framework of data as a resource for the community, interested groups and the local and national authorities to facilitate informed discussion and decision making.

The method can be described as 'action research' where the principal role of researchers is to act as a resource for the community rather than seeking to impose a pre-set solution or pre-set agenda. Such a methodology can respond to the expressed concerns of the local people and provide an effective response to them by channelling appropriate questions and focusing the issues to be addressed in discussion with other bodies.

The approach which was used in four key areas involved:

(1) A small meeting with a representative group of individuals was used to give an in depth briefing on the aims and objectives of the study.
(2) A large open public meeting at which people were arranged in groups of 3–8 to discuss and record their views on the strengths, weaknesses, opportunities and threats of the area. All answers were then collected and displayed so that they could be discussed and prioritised by everybody present. The research team then summarised the possibilities of obtaining EC and State funding through current schemes including the EC LEADER programme. Finally the meeting served as an opportunity to establish a development committee in areas where none existed.
(3) A house to house survey using a carefully prepared questionnaire.
(4) A second public meeting at which representatives of statutory organisations spoke on their work in the area and a community representative summarised problems and opportunities. This was followed by an open discussion session.

Principal recommendations

The management of recreational use and housing development were identified as the principal environmental issues in the Dublin/Wicklow Uplands. Land use changes due to reclamation, commercial forestry, peat-cutting and grazing, which were also analysed in this report are of less significance. These pressures are focused on an essentially rural community which is continuing to witness a decline in farming, its principal economic activity. There is an urgent need to provide job opportunities based on a strategy of sustainable development, and for the establishment of a system of co-operative management between landowners and conservation interests to address these management issues. Future management cannot be the responsibility of any one agency and must involve the co-operation of residents and farmers. The National Park proposal as it is presently constituted is inadequate to deal with these management issues.

The current implementation of the LEADER programme for integrated rural development in County Wicklow, the interest of central government in establishing a National Park and the prospect of supports from the EC for environmentally friendly farming, mean that residents and all the people and organisations concerned with the management of the Uplands have a unique opportunity to agree on and implement an integrated plan to balance conservation and development. The implementation of this plan should be the responsibility of residents, landowners, existing agencies and state and local authorities all of whom should be represented on a consultative body called the Uplands Committee . Members include representatives of Wicklow Co. Council, the National Park authority (Office of Public Works), community, farmer and forestry representatives (including the LEADER company) and nominees of recreational interests. It is currently being set up under the authority of the County Council and will be supported by a small executive staff. Through open discussion, the *Uplands Committee* should allow for the co-ordinated deployment of state and local authority resources to tackle the principal issues affecting the conservation and development of the Uplands. It should promote and co-ordinate the production of small scale land use management/development plans for sub-regions in the Uplands, focusing on land use, forestry, conservation, recreation, housing and physical development. These plans should only be drawn up after widespread consultation. In their final form they should specify the measures to be taken and the responsibilities of all concerned groups to ensure that future development is balanced with conservation and that the plans take due regard of local interests. These plans should then be considered for inclusion in the County Development Plan. The Committee also has a role in seeking funding from the EC to support sustainable rural development, initiate research and promote education and training programmes.

The name suggested for this management concept which represents a mechanism for achieving sustainable development in rural areas could be COMHAR — an Irish word which can be interpreted as 'many hands on the plough'. COMHAR CILL MHANNTAIN (the Irish name for Wicklow) could then be the name for the Uplands area which is managed as described.

There are national implications for this policy in the Dublin/Wicklow Mountains which are appropriate to other areas of conservation importance. It was recommended that the government should consider establishing appropriate inter-departmental co-operation to assist the formulation of plans for sustainable development in other rural areas of key conservation importance and development potential.

The County Wicklow LEADER Plan for 'Bottom Up' Integrated Rural Development

The LEADER Programme of the European Commission was established in response to the rapidly deteriorating situation of agriculture in peripheral areas of the European Community. The programme provides up to 50% grants to rural communities who have put together an integrated programme of training and job creation through rural tourism, and local companies exploiting natural resources, or providing new services. There is a particular emphasis upon the use of Information Technology as an aid to training, communicating, and marketing local products and services. Each programme must serve as a model for Europe of practical and sustainable innovation which is initiated and developed by communities themselves with the help of existing agencies in the region.

As part of the Wicklow Uplands Study (see above), the research team were involved with drawing together community-based proposals for an integrated rural development plan which was submitted to and approved in late 1991 for funding by the EC LEADER Programme.

The programme for the Wicklow region was drawn up to fit the region's unique situation in Ireland, being a highly rural area in a landscape of outstanding scenery, but adjacent to a capital city with a population of about 1 million people. The aim of the Wicklow programme is to test strategies which might improve the social, economic and cultural stability of the rural communities of the region and which might help to bridge the gap of understanding by urban people of rural communities.

The general objectives are therefore to pioneer a system of integrated rural development which is specifically designed as a European model for areas of landscape of outstanding quality which are within 100km of an urban population of at least 1 million people. The main elements of the Wicklow programme are:

(1) A new company, Wicklow Rural Enterprises Ltd, employing 3 full-time rural development workers.
(2) Initiating the following integrated series of training programmes: (i) Rural enterprise development, (ii) Leadership, (iii) Rural business development, (iv) Rural tourism, (v) Heritage in rural tourism, (vi) Information technology in rural development, (vii) Applications of IT to marketing rural enterprises.
(3) Provision of 190 new bedrooms for tourist accommodation at farms and country houses, and three new caravan-camping parks.
(4) Provision of attractions and facilities for visitors (small heritage centres, pitch and putt/ golf courses, facilities for riding and fishing etc.)
(5) Setting up new enterprises providing new services and products.
(6) Exploiting and marketing local products based on local natural resources such as timber and farm products.
(7) Establishing a network of Information Technology Service Centres to provide international communications facilities through Videotex terminals, Minitel terminals and computer modems. In addition an interactive multimedia computer system is to be set up for marketing the visitor attractions and services of the region.

The total cost of this two year programme is Ir£4 million covering 105 projects, with a LEADER contribution of Ir£2 million, the remaining Ir£2 million coming from individual project proposers.

Design and Implementation of Interpretative Displays

Though universities by their nature are major repositories of knowledge and expertise on natural and human heritage, they usually lack expertise in designing and implementing interpretative displays which are suitable for tourism. Trinity College has a policy of encouraging links with industry through the establishment of campus companies whose business usually stems from on campus research expertise. There are now 22 such companies and one of these, Eigse Ltd, consists of a team which combines expertise in heritage research, exhibition design, graphic design, production of audio-visual displays and of films. The close link with a university helps to ensure that a high standard of heritage information is presented to visitors.

Examples from County Wexford of New Methods for Marketing Rural Tourism and for Promoting Education in Local heritage

The successful development of rural tourism, particularly of community-based tourism, is very dependent upon national and international marketing of the attractions which are unique to each area. It is difficult for individuals or community organisations to do this by themselves. It is also important that when visitors are in an area, they can browse through in some detail the places or activities which might interest them. County Wexford is particularly rich in natural and man-made heritage; however, it is difficult for visitors to find out about it. In an attempt to remedy this situation we have developed a number of new marketing and longer term educational tools.

The 'Discover' multimedia system

The 'Discover' system is a user-friendly interactive multimedia computer system, based on Apple Macintosh microcomputers operated by a touch screen. Through support from the European Space Agency, a demonstration 'Discover Wexford' system has been developed for marketing tourism based on heritage and activity resources in the county. It has been designed and produced by the Trinity College campus company ERA-Maptec Ltd. 'Discover Wexford' contains text and graphics for about 800 sites which are accessible to the public. There are about 400 colour slides in digital form and small amounts of analog video which include footage taken from a helicopter and flight simulations over the coastline generated from satellite imagery. The system allows untrained users to explore an area by displaying maps, photographic quality colour images, video, sound and text. This material can be stored on Laser Disc, Hard Disc or CD ROM; it is extracted in real time from an advanced database and displayed by touching appropriate topic icons on the screen. From these displays, the user can build up on screen detailed thematic maps which show where things of interest are. These maps can then be printed and taken away. Discover has been designed as a graphical interface to a remote database so that it can be used in a local area network, or in an international network using broadband communication technology. Real time data such as times of concerts, ferries, sporting events etc. can be typed in from a remote terminal using a modem. New colour graphic displays can also be added overnight from a remote terminal. The system records every query so that it is easy to monitor what visitors are interested in. As a result of testing the system over the last two years in tourist information offices, the design has been simplified to maximise its use by visitors who want to use it as an effective way of planning their exploration of the county. The most used element has been the colour photographic images and accompanying text information for all the accommodation in the county which has been approved by the Tourist Board. It is essential that the displays are of high quality design if the system is to be effective in attracting the attention of viewers who are used to high quality image design in television. With the establishment of international broad band telecommunication systems over the next 10 years, multimedia systems are likely to become a very important way of choosing holiday destinations.

Data collection for multimedia systems

A major difficulty in using multimedia technology comes from the cost of data collection. The approach which we use has been to assist communities in collecting their own data and to provide training for data design and entry. The collection of data must be guided by the interpretative plan for the area (see above). Depending upon the size of the area and the quantity and quality of existing data, two alternative strategies can be used in the production of a data base for a multimedia system. Firstly, data may be collected and preliminary design carried out within the training course summarised earlier. Alternatively, if funding is available, as was the case in County Wexford, it may be possible to collect data by employing a small expert team drawn from the community and from outside.

Publications

The 'Discover' multimedia system has proved to be a valuable method of allowing visitors to browse through the attractions of an area. A considerable benefit of developing such a multimedia system is that all data is in digital form and it is cheap and easy to produce or update colour brochures and maps either for general interests or which may be linked to tours or specialist interests (e.g. fishing, walking etc.).

Using the data base for the 'Discover Wexford' multimedia system, four well illustrated thematic booklets entitled *Introducing Co. Wexford; Nature and Landscape; Tracing Wexford's Past, Customs, Farming and Villages* have been produced for visitors to complement in paper form the Discover system. They have been published and are distributed by the EC LEADER company for County Wexford (WORD).

If heritage based rural tourism is to be sustainable, it is essential for young people to become well informed about the heritage of their home areas. In an attempt to promote this aspect of education in this area, we have produced for WORD a gazetteer for County Wexford which contains descriptions of every heritage site in the county. This is intended to be a source book which is available in libraries. The information in the gazetteer has also been reorganised into a series of 48 booklets covering the heritage of every parish in the county. Our aim has been to assist every primary and secondary school in the county to use the heritage adjacent to them as a teaching resource. From experience gained from teachers using these parish heritage booklets, we plan to produce a series of heritage teaching manuals.

Conclusions

The development of heritage and activity based rural tourism requires very careful planning and networking if it is to be sustainable. The following measures are recommended:

(1) Production of a resources data base for the area.

(2) Production of a data base and report summarising the existing tourism product in the area.
(3) Production of an interpretative plan.
(4) Delivering a training programme on heritage in rural tourism.
(5) Production of an integrated development strategy which includes rural tourism.
(6) Production of digitally based marketing materials such as interactive computer multimedia systems, thematic guide books, brochures and maps.
(7) Production of source books to support and encourage the use of local heritage in primary and secondary level education in rural areas.

Acknowledgements

We are grateful to Wicklow County Council for allowing us to refer to some of our work which was commissioned by them. In our studies in Counties Wicklow and Wexford we owe a debt of gratitude to the following for their contributions: Graham Barrow and Jillian Binks (CEI Manchester Polytechnic); Tom Broderick and Blaise Treacy (Wicklow County Council); Jonathan Blackwell, Ned Culleton, Brendan Culleton, Patrick Daly, Zena Hoctor and Brian Madden (Natural Resources Development Centre); Fred Aalen, P.J. Drudy, John Jackson, David Jeffrey, Hilary Tovey (Trinity College).

References

Bord Failte (1992) *Proceedings of the Second Conference on the Development of Heritage Attractions in Ireland, 28–29th January 1992, Royal Hospital Kilmainham.* Bord Failte

— (1992) Developing Sustainable Tourism: Tourism Development Plan 1993–1997, a Draft Summary. December 1992. Bord Failte.

Commission of the European Community (1988) The future of rural society. *Bulletin of the European Communities,* Supplement 4/88. Brussels: CEC.

Convery, F.J. *et al.* (1989) Tourism in Co. Wicklow: Maximising its potential (pp. 1–218). A report prepared for the County Wicklow Tourism Authority.

Curtin, C. and Wilson, T.M. (1988) *Ireland from Below: Social Change and Local Communities.* Galway University Press.

Feehan, J. (ed.) (1992) *Tourism on the Farm* (Proceedings of two conferences on farm tourism in Ireland). Special publication, Environmental Institute, University College Dublin.

Keane, M. and Quinn, J. (1990) Rural development and rural tourism. Social Sciences Research Centre, University College, Galway.

Phillips, W.E.A. *et al.* (1992) An interpretative plan for County Wicklow (pp. 1–146). A report prepared for Wicklow County Council.

Tubridy, M. (1993) A plan for community led sustainable tourism in NW Connemara, 1994–1997 (Parts 1–3). A report prepared for FORUM (the Rural Development Partnership in NW Connemara), Letterfrack, Co. Galway.

Tubridy, M. *et al.* (1992) A strategy for the sustainable development of the Wicklow Uplands (pp 1–208). A report prepared for Wicklow County Council.